We all like to believe in gut feelings, but intuition is a terrible guide. If we believed what our minds tell us, we'd think the world is flat, the Sun orbits the Earth and that a day is made of 24 hours (it's not).

At some point, we humans noticed how profoundly limited and biased our senses are, and how routinely our instincts fail. And so we invented science and mathematics to help us to see the world as it really is. In their joyful and exhilarating book, Hannah Fry and Adam Rutherford take us on a journey of scientific discovery, from the origin of the universe to its inevitable ultimate demise, from the beginnings of life on Earth to the possibility of wondrous alien life forms elsewhere in the galaxies, from the darkest depths of infinity to the brightest recesses of our minds.

Along the way, they'll answer head-scratching questions like: Where did time come from? Do we have free will? Does my dog love me? And they'll share tales of great wisdom and hard work, including the many fumbles and mis-steps, blind alleys, blind luck and some *really* bad decisions that, when put together, amount to the biggest story of all: how a species of mostly bald apes, with unique and innate curiosity, decided not to be content with things as they appear, but to poke at the fabric of the universe and everything in it.

Reality is not what it seems; but if you're ready and willing to set off in search of it, this is your guide.

RUTHERFORD & FRY'S

COMPLETE GUIDE
TO ABSOLUTELY

EVERY THING*

*ABRIDGED

HANNAH FRY &
ADAM RUTHERFORD

Illustrations by Alice Roberts

BANTAM PRESS

TRANSWORLD PUBLISHERS
Penguin Random House, One Embassy Gardens,
8 Viaduct Gardens, London SW11 7BW
www.penguin.co.uk

Transworld is part of the Penguin Random House group of companies
whose addresses can be found at global.penguinrandomhouse.com

Penguin
Random House
UK

First published in Great Britain in 2021 by Bantam Press
an imprint of Transworld Publishers

A CIP catalogue record for this book
is available from the British Library.

ISBN 9781787632639

Designed and typeset in 10.5/15pt Gill Sans by Julia Lloyd.
Printed and bound in Great Britain by Clays Ltd, Elcograf S.p.A.

The authorized representative in the EEA is Penguin Random House Ireland,
Morrison Chambers, 32 Nassau Street, Dublin D02 YH68.

Penguin Random House is committed to a sustainable
future for our business, our readers and our planet. This book
is made from Forest Stewardship Council® certified paper.

MIX
Paper from
responsible sources
FSC
www.fsc.org   FSC® C018179

To the NHS, who saved both our lives
during the writing of this book

CONTENTS

INTRODUCTION

Close your eyes.

Admittedly, reading generally requires your eyes to be open. If you are holding a printed copy of this book, in a few seconds you will certainly need them open, because you can't read the rest of what we're about to say with them shut, obviously.

But for now, close your eyes.

During that brief moment of darkness, not much changed. The words stayed on the page; the book, thankfully, was still in your hands. When you opened your eyes, indeed when you opened them this morning after some restful sleep, the light flooded in, and you recognized everything as pretty much the same as when you'd closed them. Reality persists whether you are paying attention to it or not. All of this might seem very obvious. Silly, even. But this is a fact that, once upon a time, you had to learn.

Next time you're playing with a baby, try taking a toy and hiding it under a blanket in front of them. If they're less than about six months old, they won't pull the blanket away to get the toy back, however much they were enjoying playing with it beforehand. That's not because they lack the skill to grab and move the cloth – it's because, unlike you, they simply don't realize the toy still exists. To their tiny mind, it simply poufed out of existence the moment it vanished. This is why babies find the game of peekaboo such fun. It's why peekaboo is played by every culture, by all humans all around the world. When you place your hands in front of your face, a very young and immature mind assumes that you have literally disappeared, and possibly ceased to exist. The joy in discovering that your existence hasn't been erased from the universe shines out in the baby's giggles when you take your hands away.

Peekaboo exemplifies quite how badly equipped humans are for comprehending the universe, and everything in it. We're not born with an innate understanding of the world around us. We have to *learn* that stuff – including people – doesn't just vanish when we are not looking at it. In babies, it's an important milestone in development known as 'object permanence' – something that many other animals never quite manage to grasp. A crocodile can be subdued by covering its eyes. Some birds can be calmed by placing a cover over their cage. It's not just that they find the darkness soothing – they don't realize the pesky human bothering them is still there, on the other side of the cloth.

Why should their brains care about object permanence? The primary motivation of almost every organism that has ever existed has been not to die – at least, not until it has had a chance to reproduce. Most life on Earth is altogether unconcerned with the question of why things are the way they are. Dung beetles navigate at night using the Milky Way as their guide, with limited interest in the structures of galaxies, or the fact that almost all of the mass of the universe is (so far) unaccounted for.* The tiny mites that live in your eyebrows are oblivious to the concept of symbiotic commensalism whereby they innocuously feed off us. Until now, you were probably entirely unaware of them too, *but they are definitely there.* A peahen has no interest in processing the complex equations that explain *why* she finds that ridiculous tail on a peacock so irresistibly sexy; she just kinda likes it.

Only one animal has ever asked these questions – us. At some time in the past hundred thousand years or thereabouts, some mostly bald apes started to get curious about pretty much everything. The brains of these apes had been growing bigger over the previous million years or so, and they began doing things that no other animal before them had done. They started drawing, and painting, and making music, and playing peekaboo.

It's important to not get too mushy about this. Prehistoric life was still pretty wretched compared to today, and survival was still everyone's primary concern. But our ancestors had taken a step away from the rest of nature by considering not just the immediate concerns of survival, but the whole universe, and their place in it. However, we are still apes – and much of our brains and bodies is still fundamentally concerned with just

---

* To discover this, scientists put little hats on dung beetles at night and watched them as they got totally lost. Not everything in science has to be high-tech or complex. Sometimes it's just putting a hat on an insect.

living and reproducing. Physically, and genetically, we haven't changed much in the last quarter of a million years. Take a woman or man from Africa 300,000 years ago, transport them forwards in time, tidy them up, give them a haircut and stick them in a nice dress or sports casual, and you wouldn't be able to pick them out of a crowd today. Much of our biological hardware is largely unchanged from a time when none of these highfalutin ideas about how the universe works were of much concern to anyone.

What all this means is that our senses routinely let us down. We jump at quick, unexpected movements, despite no longer having to worry about predators trying to eat us every day. We crave sweet, salty and fatty foods – a perfectly sensible hunter-gatherer strategy, helping us prioritize high-calorie inputs when food was scarce, but much less useful when there's the option of ice cream after every cheeseburger.

These evolutionary hangovers go beyond our instincts; they affect our intuition too. If you'd asked our uneducated ancestors about the shape of the Earth, they might well have told you it was flat. It makes sense that it's flat. It looks pretty flat – and surely, if it wasn't flat we'd fall off. But it's not even remotely flat. In Chapter 3, we'll intimately explore our lumpy rock and determine that not only is it not flat, but it's not even a sphere: owing to its rotation, the Earth is an *oblate spheroid* – essentially, a slightly deflated ball that is a bit flat at the poles and a bit fat around the middle.

From our perspective, the Sun looks very much like it revolves around the Earth: every day for the past 4.54 billion years, it has come up in the morning over here, scooted across the sky and gone down over there. But in reality, the Earth orbits the Sun – and it doesn't do that in a perfect circle, either.

As far as we are concerned, the Sun is static in space while we whizz round it. But in reality the Sun and our whole solar system are charging round a point at the centre of the Milky Way at a bracing 514,000 miles per hour, completing a full orbit once every galactic year (that's 250 million Earth years). None of us have the slightest experience of that while we're sitting reading in a deckchair.

Curiosity might have marked humans as different from other creatures, but curiosity alone is not enough. When humans ask curious questions about the mysteries of reality, we don't necessarily come up with the right answers instantaneously; there's no end to the myths that we concocted to explain the inexplicable nature of nature. Vikings decided that the deafening sound of thunder was Thor charging across the sky in his goat-powered chariot, and his fearsome hammer Mjölnir was the source of lightning.* The Gunai, indigenous to Australia, thought that the Southern Lights, what we call the *Aurora Australis*, were bush fires in the spirit world.

Stories of gods, goats and ghosts are still widely believed by billions of folk on every continent. Some of those stories might be easy to mock, but intuitively they make sense, and intuition is an incredibly powerful thing. We cannot help but see the universe through human-tinted glasses. In fact, though, much is not as it seems. As you'll discover in this book, a day is not 24 hours. A year is not 365 (and a quarter) days. When we admire our star sitting just above the horizon in a beautiful sunset, it's actually already *beneath* the horizon: the atmosphere of the Earth bends the light so we can see it even

---

* Thor would eat the two goats, Tanngrisnir and Tanngnjóstr, every night, and bring them back to life with Mjölnir. That, admittedly, is not intuitive.

after it has set. Eating sugary sweets and cake doesn't make kids go nuts at parties.* More people die by drowning in the bath than are killed by terrorists and sharks *combined* every year, but no government has introduced laws on bath-time policy (yet).

Whatever way you look at it, intuition is a terrible guide.

And at some point, we curious apes realized this. We created science and mathematics in an attempt to step out of our limited human perspective and see the world as it objectively is, not merely as we experience it. We recognized the limits of our senses, and came up with ways to expand them so we could see beyond the narrow spectrum of our vision, hear beyond the range of our ears, and measure beyond distances we could see, to the unimaginably large and infinitesimally small.

Since then, we've been striving to learn how reality truly is. That is what science is. We have been doing this for hundreds, if not thousands of years, but not always successfully. Earlier attempts are often easy to poke fun at, and not always far from the gods and goats. Plato believed that we could see thanks to invisible ray beams shooting from our eyeballs, probing and investigating everything they touched; but then again, he didn't have theories of the electromagnetic spectrum or neuronal phototransduction. Early biologists thought that sperm contained a homunculus, a teeny tiny version of a person, and

* The best data we have suggest that children just go nuts at parties, whatever they eat. In tests, when scientists gave children food with no sugar, but told the parents there was sugar in their drinks and cakes, the parents rated the behaviour of the kids as worse, when in fact it was no different. In this sense, parents behave worse at parties than kids, who are just being kids at parties.

that the woman's job was merely to act as a vessel to incubate this mini-person until it was a full-size baby. Isaac Newton was an alchemist who put much more effort into trying to turn lead into gold than into his work on the mechanics of the cosmos. Galileo was an astrologer as well as an astronomer, and did horoscopes for paying customers when he was short of cash. Van Helmont, the father of gaseous chemistry, believed that mice would ping into existence if you just stuffed a vase with some wheat seeds and a sweaty shirt and left it in a dank basement for 21 days.

Science has got an awful lot wrong over the years. One could argue that it is, in fact, science's job to get things wrong, as that is the place from which you can start to be less wrong, and after a few rounds get things right. On the whole, the arc of history curves in a progressive direction. We've built huge civilizations that lasted for centuries. We've changed nature, and bred animals and crops that feed billions. We've used maths and engineering to put up buildings that last for millennia, and to construct ships that allowed us to traverse the globe (and in doing so, affirm that it *is* a globe). We've created spaceships that can master the dynamics of the solar system and visit alien worlds billions of miles away. We've even populated an entire planet with robots. Some day soon, one of us will embrace all the brilliance of the people who came before them, and will set foot on that planet and become the first ape on Mars.

All of that is worth celebrating. Science and maths are a toolkit, the ultimate shed, crammed full of the most wonderful instruments and ideas, devices and gizmos to augment our abilities and expand our senses so that we can observe more and more of reality.

This book is a guide to how we have tried to suppress our

monkey-brains to see the universe as it really is, rather than how we perceive it to be. It's about the difference between what feels intuitively true and what scientists have discovered to be the truth. Often, that truth is far harder to believe.

Your mostly bald ape guides come from very different domains of science. Hannah is a mathematician, who specializes in crunching colossal swathes of data to understand patterns in human behaviour. Adam is a geneticist, who squints at DNA to see how living things adapt and survive, and how life on Earth evolved in all its magnificent splendour. Like everyone else in all domains of science, we are merely trying to figure out how stuff works. The mistake that people sometimes make, and often teach, is that science is a bank of knowledge. After all, the word 'science' comes from the Latin *scire* – to know. But science isn't just about knowing; it's about *not* knowing and having a way to find out.

Here, in this book, are answers to questions that at first may seem simple, silly or utterly baffling. What would aliens look like? Does my dog love me? And what does a space death cult that is utterly dedicated to a forthcoming apocalypse do when the apocalypse doesn't show up?

The questions themselves are straightforward enough (maybe not the space death cult one); but in working out the answers, we find that they accidentally spill the real secrets of the universe, the ones we can only see when we switch off our monkey-brains and engage the tools we invented to get past our evolutionary blockades. They're questions whose answers reveal how little we can really trust our instincts, and how far beyond ourselves science has armed us to venture.

This book features tales of the universe and how we have tried to understand it – all the big concepts such as time, space, spacetime and infinity, and questions such as: What time is it? Not as in 'It's way past my bedtime' or 'Isn't it time you took that library book back?' What is the actual, universal, unequivocal, absolute measure of the way that the present sits somehow between stuff that's already happened and stuff that is going to happen? The answer will take us on a trip among imperilled sailors, anxious bankers, ancient coral, Einstein and space lasers. But we'll also tell stories of why we humans are so prone to getting things wrong, and how we get round that. It's about how evolution has furnished us with wonderful senses that can and do deceive us, but also with brains that allow us to revel in the wonders of the universe and bypass all this baggage we carry around in our heads.

This is a book of our favourite stories – tales of how we know the things we know, tales of our fumbles and mis-steps along the path of ever-expanding knowledge. The errors, egos, insight, wisdom and prejudice of scientists and explorers; the hard work, tragedy, blind alleys, blind luck and some really, *really* bad decisions – all these are bits of the jigsaw of our history that has brought us to where we are today. This book is a celebration of how being wrong is the way to get things right; of how changing your mind isn't always easy, but how being prepared to do so is a virtue (in general, but particularly in science). It's a journey through time and space, and through our bodies and brains, showing how our incredibly potent emotions shape our view of reality, and how our minds tell us lies. Put together, these tales amount to the biggest story of all: how a species of mostly bald ape, with its unique and innate curiosity, decided not just to be content with things as they appear, but to poke at the fabric of the universe and everything in it.

Reality is not what it seems; but if you're ready and willing to set off in search of it, this is your guide to how the best tools ever invented let us see things as they really are.

CHAPTER 1

ENDLESS POSSIBILITIES

A stale but not unpleasant smell fills the air. The ceiling hangs low, tempting you to touch it with your outstretched fingers. Along four of the six walls that surround you are rows of leather-bound books, dusty creased pages and ancient ink that hasn't seen sunlight for years, maybe centuries.

Your room is not unique. Through small ventilation shafts, there are glimpses of other galleries, above and below, one after the other, endlessly into the distance. Hallways on your own floor, reached through doors in the other two walls, spin off into other hexagonal galleries, each identical to the room you are in. Galleries brimming with books, each book brimming with words.

This is not a normal library. You're standing somewhere in an unfathomably vast honeycomb, a labyrinth of the written word. Somewhere within these walls is a copy of every book ever written, and every conceivable book that will – or could – ever exist. Forget for a moment these humble pages you are reading. This library is the *true* guide to absolutely everything.

This is the Library of Babel, a figment of the literary imagination, created by Argentinian author Jorge Luis Borges. It is the centrepiece of a short story of the same name written in 1941 about a universe where every possible thing was committed to paper – a story that plays with a single idea. If, somehow, you had access to absolutely everything, how much could you possibly know?

Absolutely everything

The books in Borges' unimaginably vast library are built from every possible combination of letters, spaces, commas and full stops – all those that form words and sentences, and many more that don't. This library holds every word that anyone has spoken, thought or written – and will ever speak, think or write – in every conceivable order, and all the random arrangements of meaningless gibberish in between. In Borges' own words, hiding on the shelves of his library you will find:

> **the detailed history of the future, Aeschylus' *The Egyptians*, the secret and true name of Rome, my dreams and half-dreams at dawn on August 14, 1934, the proof of Pierre Fermat's theorem, the complete catalogue of the Library, and the proof of the inaccuracy of that catalogue.**

It's a fantastical idea. But the Babel library isn't just a figment of Borges' imagination. Someone has built it.

At least, a version of it. In 2015, Jonathan Basile, a student at Emory University in Atlanta, Georgia, built the Babel library – in digital form, with a few necessary practical constraints.

Imagine for a moment a library of pages each containing words of only five characters. It would be easy enough (although not a huge amount of fun) to write out the combinations yourself:

aaaaa

aaaab

aaaac

. . .

and so on. You would run out of ink fairly quickly. Print out the full list of combinations for five characters in 12-point type, and your sheet of paper would need to be around 60 miles long.

And that's just the combinations for five letters, not the 410-page tomes that Borges imagined. Jonathan Basile quite quickly realized that running sequentially through everything just wasn't going to be possible. Aside from taking for ever to compile, a digital library built letter by letter would need so much storage space that even if the entire observable universe were filled top to bottom with back-to-back hard drives, it *still* wouldn't be enough to see Borges' dream realized.

Basile was going to need a short cut. First, he decided to limit his library to contain only every possible *page* of writing, rather than every possible book. It's still a ridiculous endeavour: every possible page of 3,200 characters built from 26 letters, along with spaces, commas and full stops. But it's marginally more achievable.*

Then he came up with a very clever idea to spare him from spending his next gazillion lifetimes typing out the entire library.

Like Borges', Basile's infinite library is arranged into virtual hexagons – four walls of books (and two doors to the adjacent

---

* Basile's algorithm can convert the page numbers, in base 10, into a 3,200-digit random number in base 29. It does so by using the page number as the starting point in a carefully crafted pseudo-random number generator. Each digit in base 29 relates to a letter of the Roman alphabet, plus spaces, commas and full stops, so it's a straight swap between the gigantic random number generated by the algorithm and the page of text. Basile has also made sure his algorithm spits out *every* possible output exactly once and once only, so that *every* possible page can be found somewhere in the library. And – even more ingenious – his algorithm is reversible, meaning that if you give it a block of text, it can convert that to a number in base 29 and run backwards to give you the number of the page on which it appears – essentially making his library searchable.

rooms), then shelves, volumes and pages. The pages inside are all organized to give each of them its own unique reference. For instance, here's the first line from Hexagon A, Wall 3, Shelf 4, Volume 26, Page 307:

```
pvezicayz.flbjxdaaylquxetwhxeypo,e,tuziudwu,
rcbdnhvsuedclbvgub,sthscevzjn.dvwc
```

Not one of the library's great thrillers, admittedly.

It's the relationship between the reference number and the text it identifies that made the library possible, though. Basile's trick was to use that unique reference number to create a code that could only be deciphered in one way: an algorithm that could reliably generate a unique page of text from a unique reference number, when asked.

You can read a bit more about how Basile's algorithm worked in the footnotes,[*] but the important point is this: every page number within the library is indelibly linked to a single page of text. Give the algorithm a reference number and it will tell you what's on that page. Give the algorithm a page of text and it will tell you the reference number.

It's the algorithm, not the librarian, that does all the heavy lifting. Without anyone having to type anything out, every page

---

[*] Basile's library contains every combination of all 26 letters of the English alphabet, a to z. The library is not in English, per se. Any word or string of words written in any language that makes use of the Roman alphabet can be found within its pages. On that point, Basile's library does differ slightly from Borges' original story, which used every combination of only 22 letters. Borges got to 22 characters by starting with the modern Spanish alphabet of 30 letters, and omitting all the double letters (ch, ll, rr) as well as ñ. He also ditched w, q, k and x as unnecessary (a questionable decision), giving him a total of 25 characters once you include spaces, commas and full stops.

is already predetermined – preordained, even – and simply summoned by the algorithm. Every page already exists, just waiting for someone to pull it from the shelf.

The first paragraph of this chapter is in there. If you don't believe us, it's sitting in the hexagon with reference ending 993qh, on Wall 3, Shelf 4, Volume 20, Page 352. We didn't put it there; it was there already:

```
lmgumfkwwomyzzoxpj,qyoynhdaqhtslvacnaicu
varzkdjzzazvmppap  bteq  ezlblbsjjaesejhtz
vv.b,uc.ofrx.ul gidtfhqpwikgygk,kvq. rosf.
bgdeurubwp,eqns.huyiyrnz.cocddh q.,,znuav.
wvqwwcwohn chmrwua stale but not unpleasant
smell fills your nostrils. the ceiling hangs
low, tempting you to touch it with your
outstretched fingers. along four of the six
walls that surround you are rows of leat
herbound books, dusty creased pages and a
ncient ink that hasnt seen sunlight for
years, maybe centuries.foxvpx.krv,.pwsmwv
iuyuhkdrcx,,wplknvo,dsopqcrmhduenco   rnpb
vdwd.xxxgsareodhjnjzf.xsxkf,aaofbmvcqlzlk
ktkweib.xhc.r,pbfkdcxhsznrjocvlaqvbn.,j.
```

We spent some time crafting that paragraph, so it is slightly galling to see it already written, effortlessly, by an anonymous bit of code. Of course, it is pointless to be annoyed at infinity. There's a page in Basile's virtual library, just waiting for you to find it, that's made up entirely of spaces, except for your name written squarely in the middle. There's a page with the story of your day today. There's a page with the name of your first true love and how you met, and one describing you murdering your current partner with a ladle. There's a page with a beautifully written story about a dog called Molly, and one spelling out precisely how you will die. There are others containing every possible story that could be written about you, but with one detail slightly off, like every possible mis-spelling of your name – and also in French, German, Creole, Italian and every other language written in the Roman alphabet. There is a website, in short, that already exists, and contains the sum total of all human knowledge.

THE ARECIBO MESSAGE

Creating codes that can only be deciphered in one way is a common theme in maths. In 1974, two people, on behalf of all humanity, used one of these codes to attempt contact with alien life forms.

Contact with aliens will be arguably the single most important event in the history of humankind. The question is, what should our first statement be? Anyone who's ever endured a networking drinks reception will know that deciding what to say when walking into a medium-sized room with people who share the same interests as you is terrifying enough. So what should the message be – global in its scope, intergalactic in its ambition – when announcing to the residents of the universe that we are here?

Astrophysicists Frank Drake and Carl Sagan had an idea. They concocted an ingenious encoded message from all humankind, and on 16 November 1974 had it beamed in FM from the giant radio telescope in Arecibo, Puerto Rico.

What did they say? It's hard enough to get a dog to understand English, let alone announce your species to an alien civilization. But Drake and Sagan were clever sausages, and they figured that there are ways of encoding messages using the universality of maths. Prime numbers are indivisible by every number except one and themselves. That's true on Earth, on Saturn and on as yet undiscovered planets in the Horsehead Nebula; and so Drake and Sagan used prime numbers to encode their message. The broadcast consisted of 1,679 binary bits. They figured that a civilization smart enough to pick up the signal must have an understanding of maths sophisticated enough to recognize that 1,679 is a semiprime, which means that it can only be divided by two prime numbers – 23 and 73.

Picture the scene. An alien astronomer picks up this strange signal from deep space. After some pondering, it clocks that it has 1,679 bits to it, scratches one of its multiple heads for a while and then figures that the next thing to do is lay out the bits in a 23 by 73 grid. And by Grabthar's hammer! A picture materializes before its 17 eyes.

On it, the alien astronomer would see images of our solar system, with a human figure highlighting the third planet from its star, our home. They would see the atomic numbers of hydrogen, carbon, nitrogen, oxygen and phosphorus, which make up DNA, also represented by a flattened double helix. And they would see a representation of the number 4.3 billion – the Earth's population in 1974. The alien would see this all-wondrous explanation of a civilized civilization alien to itself, and would immediately, urgently contact its leaders, at which point (if science fiction has taught us anything) envoys would be sent to make friends (or possibly destroy us).

This all sounds like a monumental moment. There are just a few caveats worth mentioning. Nothing major. First, it wasn't *really* an announcement to the universe, as that kind of transmission would require more energy than is available on Earth. So, more practically, it was aimed at a cluster of stars at the edge of the Milky Way – so more like shining a torch at a model village in another country.

Tiny caveat number two: here's the actual picture.

 Can you see the solar system there? The double helix? The chemical structure of phosphate? No? Neither can we. A for effort, gentlemen, F for the 1970s graphics.

Ironically, it looks more like a 1970s Space Invaders screenshot than a missive from a strange new world. The human figure was scaled to represent the average height of an American man – some telling 1970s-style interplanetary myopia there, as half of all Americans are on average 5.5 inches shorter than that. And it was out of date almost immediately. Pluto is on there as the blob on the right (obviously). Then it was the ninth planet, but in 2006 it was relegated to the status of 'dwarf planet' and booted out of the solar system's litany of planets. Maybe if the aliens did get it, and decoded the message, they thought: 'This picture is bonkers, let's give these guys a wide berth.'

And to round it all off, by the time it arrives at its destination, 21,000 light years away, *in 21,000 years' time*, the target stars won't actually be in that location. And we'll all be dead.

In summary: neat idea, poor execution.

Too much knowledge is a dangerous thing

The Babel library is quite a collection – *the* collection, in fact. It isn't the only total library, though. Jonathan Basile has also created a library that contains every possible combination of pixels. It's hard to search, but somewhere within the eternal picture library is a photo of you scoring a penalty on the surface of Enceladus with a giant iguana in goal and Han Solo, Lizzo and Charles Darwin in the field, flanked by Marie Curie in an inflatable T-Rex suit, a lion in a Marie Curie suit and George Clooney wearing false eyelashes and nothing else. This exists.

You might think that having access to the totality of human knowledge would be a good thing. A cure for all known cancers is in there, so just go in and grab it. Paradoxically, though, what complete guides to absolutely everything give you is actually very little indeed.

Borges' original story is really about generations of librarians, once filled with optimism at having the answers to everything to hand, who have slowly been driven mad by the realization that having *everything* is rather more of a curse than a blessing. All knowledge might be in there, hiding within the pages, but finding it is another matter. The signals are drowned in oceans of noise.

Think back for a moment to the example of every possible combination of five letters. Written out, they would stretch across 60 miles, but 99.91 per cent of that distance would be covered in gibberish. The real words, written one a line, would take up no more than 260 pages. Roughly speaking, that's like scattering the pages of this book along the route from Swansea to Bristol. Please don't, though. Littering is very rude.

Far from being the repository of all human knowledge, these libraries are an unimaginable and total mess. Hold on to your sanity and take a look for yourself.[*] When you flick through Jonathan Basile's library, all you'll find, on page after page after page, are random letters of incomprehensible twaddle that don't even form so much as a single coherent word. Borges writes of a legend, whispered among the librarians, about a man who came across a book five hundred years ago containing almost two pages of readable text. By contrast, the longest legible word that Basile has found, in all his looking, is 'dog'.

If you clicked through the books in the Babel library at a rate of one per second, it would take you about 10^{4668} years to get to the end. Unfortunately, the Earth will be consumed by the Sun in less than 10^{10} years (as discussed in Chapter 7, without a hint of doom) – so good luck with that.

And then, even in the vanishingly unlikely event you found anything comprehensible, how would you know if it was true? All the pages containing the cures for cancer, or stories of your death, are indistinguishable from the overwhelming number of pages that seem plausible, but are wrong by a single critical detail. There is a strange, counter-intuitive conclusion to all of this. A library containing all possible knowledge might as well contain no knowledge at all.

[*] Visit https://libraryofbabel.info/.

The circle of knowledge

Complete libraries don't have to be built from letters or pixels; they can be made from numbers, too. Consider the poster child of maths: *pi*, or π as it's usually written. It's an irrational number, which means it cannot be represented as a fraction. Its digits, 3.14159 . . . and so on, will carry on for ever without any repeating pattern. As far as we can tell, once you're past the decimal point, every subsequent digit is just as likely to pop up as any other.* If you pluck a number at random from somewhere within that infinite list, it's got just as much chance of being a 0 as it has of being a 1, or a 2, or a 3, or a 4 and so on.

It looks like the same is true for strings of numbers, too. Pull out two neighbouring digits at random from π and you're just as likely to find the number 15 as 21. Or 03. Or 58. Pull out three and 876 is just as likely to appear as 420, 999, 124 or 753.

If every string has an equal chance of appearing and you carry on going for ever, then every possible string of numbers *must* appear somewhere eventually, at least once. There is a Babel library of numbers hiding beyond the decimal point in π.

It's relatively easy to turn those numbers into text: one way is to just set A = 01, B = 02 and so on,† which leads to a rather

---

* There's an important caveat here. All this only works if π is what's known as a 'normal' number: that is, if the digits 0, 1, 2, 3 and so on each appear with the same frequency beyond the decimal point, and every combination of digits is also equally likely to come up. There are no hints that π is not a normal number (and people have checked into the trillions of digits), but no one knows for sure one way or the other. Mathematicians like to be really *really* sure before they put their money on something.
† A note for the maths fans: a neater way would be to use Basile's trick and rewrite π in base 29, before switching to the alphabet plus commas, spaces and full stops. For this to work, π would also need to be a normal number in base 29.

extraordinary conclusion. It means that π contains all the text within Basile's Babel library, and more. All text of any length whatsoever: the complete works of Shakespeare, your internet passwords, a detailed description of that thing that you really don't want to be found out for: they're all in there. Except, unfortunately, just like the Babel library, everything else is in there too. Infinity holds a shining promise of an exhaustive catalogue of everything – overwhelmed by an unending swamp of despair.

THE TOWER OF BABEL

Borges' library is named after the Tower of Babel, a mythical building in a story from the Book of Genesis in the Bible that attempts to explain why the peoples of the world speak in different languages.

Humans, the story goes, were once united by a single language. After the great flood of Noah, they migrated west to a place called Shinar, where they decided to build a tower so tall that it reached heaven. Alas, God was decidedly unimpressed by this idea, which He thought smacked of hubris, and decided to punish the overambitious human race. According to His own word, His next move was to 'confound their language, that they may not understand one another's speech', which seems a bit mean, but explains why we speak different languages.

It's one of those Bible stories (and similar versions exist in many cultures) that attempts to explain an observed phenomenon, which of course is exactly what science also does. The real story of human language evolution is crazily complicated and nigh on impossible to comprehend fully, because the spoken word leaves no fossils. We know that humans have been anatomically capable of speech for hundreds of thousands of years, and that our ancestors the Neanderthals were too. And how do we know this? It's all down to some intricate anatomy centred on one bone in the neck, called the hyoid – a little horseshoe-shaped bone under your chin that moves up and down when you swallow. The muscular attachments to this bone are wickedly intricate, and this sophistication is what makes speech possible. The hyoids of chimps, gorillas and orang-utans are far simpler, and we know that none of those guys speak, at least not when we're listening.

But the anatomical capability for speech is different from the evolution of language itself. There have been many attempts to understand the way languages evolve; some people have suggested that there were early forms, now lost in time, which were the ancestors of the words we speak today – ghost tongues such as proto-Uralic, about 7,000 years ago, the theoretical forebear of Hungarian and Sami; or a proto-Indo-European mother tongue some 6,000 years ago, which gave rise to a whole suite of languages, from Hindi to English to Portuguese to Urdu. Some have even speculated that there was one common proto-language that preceded all others – a single trunk to the language tree, rather like in the time before the biblical Tower of Babel – but most scientists now reject this idea, on the basis that humans were too diverse and too widespread to have shared one ancestral language. To the best of our current knowledge, humans have invented language more than once.

Today, languages are a beautiful quagmire of sponges, messy, ever changing, absorbing words and phrases from every culture that touches them. This is perfectly exemplified by English, which is an absurd hybrid of elements acquired over millennia from every Tom, Dinesh and Helga who invaded, migrated, took a husband or wife, traded or stole loot, and generally shared their stuff. That previous sentence, by the way, contains words derived from Viking, Latin, German, Indian, French . . . and a whole lot more.

For those not biblically inclined, perhaps the most perceptive story about language comes from Douglas Adams's *The Hitchhiker's Guide to the Galaxy*. The Babel Fish (not to be confused with the very real barbel fish) is a small, yellow leechy creature that feeds off brainwaves, and specifically those generated by speech centres in the brain. If you pop one in your ear it suckers your eardrum, and instantly translates all and any languages. Alas, the *Guide* notes, by removing 'all barriers to communication between different races and cultures, [the Babel Fish] has caused more and bloodier wars than anything else in the history of creation'.

In 2018, Google announced they'd made their own version – known as pixel buds – that worked with their translation algorithms to offer near-real-time translation. Fortunately, not enough people have bought them yet for us to find out whether Douglas Adams was right.

The Junkyard Tornado

And so the infinite library is a useless tool, a busted flush. Endless possibilities mean zero potential in the real world.

But in the twentieth century, some scientists wondered whether the library was closer to the real world than it might initially appear. Trade letters of the English alphabet for the letters that encode all life on Earth. DNA – the genetic code for all life that exists – is, after all, an alphabet of just four chemical bases, known by the letters A, T, C and G. Combine them together in different orders and you could have a basic recipe for a banana, an oyster, a winged anteater and every other possible living thing.

The question is, in a library that was built from these letters, what would be the chance of taking a book off a shelf, flicking to a page and finding the full working code for an eye – let alone an aerial vermilingua?

This was the kind of argument made by the eminent astrophysicist Fred Hoyle, who was unpersuaded by the idea that evolution was a result of random mutations. Surely the likelihood of randomly settling on the right permutations to chance upon even a single working protein – let alone one that could serve an intricate and delicate function like carrying oxygen in the blood or transforming light into energy – was infinitesimally small? In his words:

The chance that higher life forms might have emerged in this way is comparable to the chance that a tornado sweeping through a junkyard might assemble a Boeing 747 from the materials therein.

Hoyle's argument, catchily known as the Junkyard Tornado, sees life on Earth as a kind of Babel library. How could evolution have picked out a working gene from an infinite library of base combinations? Just as Basile could find no more intelligible text in his library than the single word 'dog', the chances of the genes for keratin or haemoglobin emerging are infinitesimal.

It's worth saying that, while Hoyle might not have liked the theory of evolution, he wasn't arguing *for* intelligent design instead. All the same, the Junkyard Tornado has become a favourite argument of creationists, who use it to assert that the probability of even one working gene emerging via the blind process of evolution is utterly implausible, and therefore a better explanation is that there was indeed a designer, an author of creation,* who built each protein for its specific purpose.

Both Hoyle and the creationists are, of course, absolutely correct. Evolution cannot work like this.

Fortunately, evolution *doesn't* work like this, and Darwin and all biologists can rest easy. Hoyle had made a fundamental mistake about the nature of evolution. The genetic code didn't bounce into being in completed form, and no biologist thinks it did. Evolution builds on what has come before, it tinkers with the tools available – tweaking a letter here and there, mostly rather meekly so as not to change something that works into something that doesn't.

This is not the same as Borges' or Basile's libraries, where every possibility is already mapped out in the countless pages. Genomes are books that are built up, step by step, with

---

* Technically, Adam is the author of creation, as he once wrote a book called *Creation*.

everything that doesn't work being thrown out along the way. It's a process that results in pages that are not random: pages that have been edited and curated and are full of meaning.

We can do some simple evolution with short words: evolving a dog into a wolf in six steps.

DOG
LOG
LOO
WOO
WOOF
WOLF

Every one of those is a real word, and every step of evolution has to result in an organism that can survive in the real world. All the way along, there were plenty of dead ends. We tried various steps that didn't make real words – SOG, KOO, WOOJ – but we ignored them, threw them out, and kept going until a real word emerged. We selected words that worked and discarded the rest.

That is much more like how evolution actually works. We don't really know what the first gene was, right at the origin of life; only that it reproduced, and it reproduced imperfectly. Since that point some 4 billion years ago, the same process has been going on continuously in every cell, every time a gene copies itself, sometimes with errors. When the errors generate duds, nature selects against them and tosses them out, because they make the host organism less healthy, less sexy – or, even more unhealthily, actually dead. When the errors are novel, not a problem, or are even useful, nature selects them for survival. This is why it's called evolution by natural selection.

The library of all possible genes contains everything that evolution has thrown away, and millions of things that evolution never bothered to try in the first place. The reality of the living world is that nature is a much more efficient librarian than a random whirlwind. Nature is a curator.

On the subject of curation, in another altogether less practical version of the Babel library, an infinite number of monkeys sit at typewriters; here, sooner or later, one will type out *Hamlet* and indeed the rest of the complete works of Shakespeare. In 2003 some researchers attempted a version of this experiment – admittedly scaled down, as an infinite number of monkeys would require an awkward and inevitably unsuccessful meeting with the ethics committee. For one month, six macaques called Elmo, Gum, Heather, Mistletoe, Rowan and Holly were given access to a bunch of typewriters. They produced five pages mostly featuring the letter 's', but spent most of the time hitting their keyboards with stones and forcing their own faeces into the gaps between the keys.

Like the Shakespearean macaque project, any version of an infinite library is going to involve a lot of crap. The title of the collection of science stories sitting in your hand says that it contains absolutely everything. But we are not monkeys; we are the librarians, and herein we've already, carefully and lovingly, selected the best stories for you.

CHAPTER 2

LIFE, THE UNIVERSE AND EVERYTHING

Life didn't whisk itself into being like a twister in a scrapheap. Nor did it spin itself out, trying every conceivable possibility before settling on what worked best. Every living thing on this floating space-rock beneath our feet came into being by the slow, meandering trial and error (or, more accurately, error and trial) of evolution that has been unfolding on Earth for the last 4 billion years.

Ours is not the only space-rock in town, though. There are eight planets in our solar system (it used to be nine, until poor old Pluto got demoted in 2006), along with several dwarf planets (of which Pluto is now one*) and hundreds

---

* Pluto was demoted from planetary status after a number of other similar-sized celestial objects were discovered in the Kuiper Belt, a region in the far reaches of the solar system where billions of bits and bobs of various sizes

of moons. And although there are very specific constraints – liquid, atmospheres, protection from the Sun's radiation – which limit the potential candidates for a celestial place that hosts life, there are still some juicy contenders among them. Titan – Saturn's biggest moon – has its own thick, nitrogen-rich atmosphere, complete with fluffy clouds and seasonal storms (though the rains are petrol, and the snow pure soot). Ganymede – Jupiter's biggest moon – has a liquid iron core, which sloshes around generating its own magnetic field (the Earth's magnetic field forms a planetary shield against the brutal rays of the Sun, which otherwise would shred every strand of DNA and burn up every living soul). Europa – another Jovian satellite – has a liquid water ocean just beneath the surface, rich in salts and other ingredients that turned chemistry into biochemistry on our world.

Outside our little corner of the universe there are other candidates, too. In the 1990s, the first exoplanets – planets beyond the solar system – were discovered, and since then thousands more have been verified. Millions more are awaiting classification. The Earth may be bursting with life, but the universe is bursting with planets.

The chances of there being life beyond Earth are impossible to calculate, because as yet we still only have a sample size

orbit the Sun from a long way away. The decision was made that either all the substantial new discoveries in this zone would have to be made planets, or Pluto had to get the chop. Pluto was only discovered in 1930, meaning it didn't manage even a single orbit of the Sun during its brief period of exalted status. Don't feel too sad, though – Pluto has a companion. It forms part of a binary system with its similarly sized moon, Charon (pronounced Sharon, we think), the two of them locked spinning around one another in a perpetual waltz. Plus, it's got ice volcanoes, putting it very squarely up there as a strong contender for most awesome place in the universe.

of one. But, fundamentally, this is a question with only two possible answers: either there is life elsewhere in the universe, or we are alone. From a scientific point of view, this is a win-win: either answer would be astonishing.

Most life on Earth is bacteria, tiny single-celled organisms that outweigh and outnumber every other life form. They even outnumber us on our own bodies, with each of us hosting many more bacterial cells than human ones. As they are much smaller than our own cells, by mass we are still mostly human, but by number we are mostly something else. Bacteria have been around since the dawn of life and will continue until the end of life on Earth, long after we are extinguished. Given this bacterial dominance, we think it likely that these simpler forms of life might be good models for life beyond Earth, but while we have the utmost respect for – and, more importantly, are utterly dependent on – bacterial life forms on Earth, we have to confess that they are quite dull. Especially to look at, because they are too small to see.

The real fun is to go big, and imagine what fantastical forms alien beasts might take. There is a very definite element to science which is formalized playtime. We have licence to play around with ideas and experiment and speculate. When it comes to aliens, many of the grandest of scientists have toyed around with questions of extraterrestrial life, from Big Bang enthusiast and Junkyard Tornado fanboy Fred Hoyle to one of the DNA double helix discoverers, Francis Crick. In the modern age, cosmologists Carl Sagan, Carolyn Porco, Sara Seager, Neil deGrasse Tyson and many other astronomical titans have all carefully considered alien life, while acknowledging that we have no direct evidence for it. Yet.

So let your imagination run free! Think big, spectacular, *terrifying*. Consider what evolution has created on Earth, and now free your imagination to encompass the whole universe. The possibilities are endless!*

Close encounters of the unimaginative kind

Except that we've already established that humans are really not very good at handling endless possibilities. If we ask you to think of an alien, we reckon you'll probably conjure up one of two types of imagined being in your mind's eye:

(1) A 'Grey' – the term given to the humanoid figures of so many films: thin-bodied and smooth-skinned, big glossy eyes in giant bulbous heads, probably naked.

(2) The insectoid, roughly human-sized, phallic-headed, acid-blooded, armour-plated alien of the film *Alien* and its sequels *Aliens*, *Alien 3* and several other increasingly disappointing films with the word 'alien' in their titles.

Do an image search online for 'alien', and pretty much every picture is of one of these (notable – but really not very

* In September 2020 there was a big media hoo-ha when scientists announced the presence of phosphine in the atmosphere of Venus. Phosphine on Earth is produced only by humans and a few other creatures, and there is no known non-biological source of this simple chemical. There was a frenzy of speculation that its presence in Venus's atmosphere indicated that life exists there, but we are more circumspect. Venus is very very hot, and looks very very dead. Phosphine there is more likely to indicate that phosphine gets made on other worlds by geochemistry, not biochemistry – in ways we don't know about. But who knows? We know we don't.

different – exceptions being perhaps the over-ripe-avocado-skinned E.T. from the film *E.T.*, and the three-eyed little green men arcade toys from *Toy Story*, who are later adopted by Mr and Mrs Potato Head).

What this demonstrates is not only the influence of popular culture on our thinking but also, frankly, our spectacular lack of imagination. There's no earthly reason why an alien would look almost exactly like us, save a few exaggerations such as an extra eye, or a magic glowing finger. Just take legs: we have two. The Greys, *E.T.*, the little green men and the *Alien* aliens also have two. Yet statistically, almost no creatures on Earth have two. Most have six.* Some have dozens when they're young, and six when they're all grown up.† Lots have none.‡ Most of the big ones have four. Lots have eight. Admittedly birds have two, but their main form of locomotion is with their forelimbs (otherwise known as wings). Of all the animals on Earth, the bipedal club is basically us, ostriches and kangaroos.§

There are plenty of reasons why going about on two legs has been good for us: we can use our hands to do other tasks; we can run long distances – useful for hunting on the savannahs of Africa, where much of our evolution occurred; we can see over tall grass and therefore are better suited to spotting things that want to eat us. But there are disadvantages too: back pain

* Beetles.
† Butterflies and moths.
‡ Snakes, worms, slugs, snails, jellyfish, coral, etc.
§ A couple of caveats: lots of animals are bipedal some of the time, including all the great apes, pangolins and the spotted skunk, but we're talking about walking on two legs as the main form of locomotion (known as habitual bipedalism). In the past, of course, many dinosaurs walked on two feet, and there was, terrifyingly, even a bipedal ancestor of the crocodile. Thankfully, they're all dead now.

is common; we're not great at climbing since we can't grip with our feet, so we're far less safe in trees; and since our pelvises have evolved to be narrow to help us stay upright, that also means that childbirth is as painful as an earthquake in your soul.*

The chances of bipedal life emerging anywhere, let alone in the rest of the universe, are low. We are the anomaly here – there's no reason to expect aliens to share the same oddity.

Hollywood doesn't seem to have got the memo. We have strong suspicions that movie budgets rather than scientific accuracy are the biggest driving force behind the barely imagined aliens of the silver screen that have had so much influence on our concepts of alien life. The very first movie aliens, the Selenites, from Georges Méliès' *Le Voyage dans la Lune* (1902), had heads like melons and lobster claws, but were otherwise upright and bipedal. They looked just like humans in alien suits. Mostly because they were humans in alien suits. Then there's the alien in *Alien* (which became plural aliens in the sequel, *Aliens*), which was much more like a grotesque space-cockroach, though still roughly the size of a tall human wearing a costume. That's because – stay with us here – it was a tall human wearing a costume: the 6 foot 10 actor Bolaji Badejo in *Alien*, and (among others) the 6 foot 2 Tom Woodruff Jr in *Aliens*. The alien predator in the Arnold Schwarzenegger jungle romp *Predator*‡ (1987) had actor Kevin Peter Hall squished

* There are some who argue that the relative sizes of the mother's pelvis and her baby's skull means that human babies necessarily have a much shorter gestation period than other mammals (and are therefore born much more helpless). Leave it any longer, and maybe that famous breakfast scene in *Alien* wouldn't be quite so far-fetched after all.
‡ And its sequels *Predator 2*, *Predators* and *The Predator*. There were also two films of *Aliens vs Predator*. Hollywood *really* needs to work on its titles.

inside; E.T. in *E.T.* was Pat Bilon, plus testes-textured latex; and in *Under the Skin*, an alien played by Scarlett Johansson wears a suit made from the skin of Scarlett Johansson.

If real aliens are out there, they will be far wilder and weirder than any prosthetics the wardrobe department at 20th Century Fox can glue on to an actor. Evolution is much more imaginative than we are. It's constrained only by what works, not by what's dreamed of in our philosophy. And there's certainly no earthly reason for extraterrestrial evolution to insist that aliens might be even roughly the same size as us.

Size matters

If size were distributed among humans in the same way as our other attributes, such as wealth, then you'd occasionally see people wandering around who were taller than the Empire State Building. Trouser-shopping would be a nightmare, especially for Bill Gates and Jeff Bezos, who'd find that the Earth's atmosphere barely covered their ankles.* As it is, we can design cars and doorways to be roughly one single size, a size that works for most humans.

Scientists, however, are doing just fine with their naming conventions: a whole genus of Brazilian goblin spiders is named after the characters and crew from the *Predator* films, including *Predatoroonops schwarzenneggeri*, *Predatoroonops peterhalli* (after Kevin Peter Hall), and *Predatoroonops mctiernani* after the director John McTiernan.
* The average US citizen's net worth is $250,000. At the time of writing, Bezos is worth more than $200 billion, which is about 800,000 times that. The average height of a human is 1.65m, so Bezos would scale up to more than 1,300km tall. The Kármán line is one commonly used boundary that separates the Earth from space, and is set at 100km.

Humans sit somewhere in the middle on the spectrum of known animal sizes. Much bigger than pigeons, far smaller than hippos. Positively gigantic when compared to ants; dwarfed by an elephant. Our galactic companions, if they're out there, could lie anywhere within that spectrum or beyond.

A life form, as we know it, has to have an inside and an outside. It has to extract energy from its surroundings in order to continue to live, and to reproduce. We can't conceive of a way that life could exist that isn't Darwinian in its evolution. But we often focus on the organism itself, rather than the organism as it exists in its ever-changing environment. Evolution is a prerequisite for life because the environment always changes, and so to live is to adapt. Life, via evolutionary change, will adapt to become as big or as small as its environment will allow, if doing so is advantageous to its survival.

On Earth, the creature that holds the undisputed crown for the largest ever to have lived isn't a dinosaur or some other prehistoric creature — it's one that is still with us today: *Balaenoptera musculus*, the blue whale. Each of these behemoths weighs up to 180 tonnes and can be as long as 30 metres — which, to put it in the age-old language of 'standard science metrics of relatable comparisons™',* is about the weight of a Boeing 737 aeroplane and the length of a basketball court.

At that size, creatures need to have their weight supported by their surroundings. Whales are slightly more dense than salty ocean water, and so would sink to the bottom if they stopped trying. (Which is just as well, or ships would have to navigate

* Human hair (width only), tennis ball, melon, basketball, small dog, turkey, large dog, VW Beetle, double-decker bus, tennis court, aircraft (specific models by choice), basketball court, football pitch, whales, Wales. This is non-negotiable. Please don't write in.

through rotting whale carcasses bobbing up and down in the waves like corks every time they tried to cross the ocean.) By inflating their lungs, whales can alter their own density and make themselves neutrally buoyant or lighter than water at a whim, giving them the impressive power of elegantly gliding through the sea without feeling the force on their gigantic frames. Watery oceans are not unique to Earth, though. There are two other spots in the solar system that could potentially house their own alien whales.

Saturn's moon Enceladus is a tenth of the size of its sister Titan, but much brighter. From space it looks like a shiny white snowball, with thick ice reflecting the heat of the Sun away to give a lunchtime temperature of a brisk −198°Celsius. But there are tectonic cracks in that icy crust, which reveal the liquid oceans underneath. Just like on Earth, where our rocky crust cracks, splits and spits out the molten interior deep beneath our feet, ice volcanoes on Enceladus spew liquid straight into space, and in 2005 the spaceship *Cassini* performed daredevil fly-bys, close enough to inhale the spray and analyse its composition. It was salty water, containing dissolved sodium chloride, hydrogen, complex carbohydrates and other very Earthly oceanic chemicals. Some of the spray from those geysers falls back to Enceladus as snow, and some of it supplies content for one of the rings of Saturn.

What this means is that, just as we have liquid rock beneath our crust, Enceladus has liquid oceans, not dissimilar to our own seas. Something in the planet core is keeping it warm enough to not freeze like the surface, but we don't know what is generating that warm glow.

The oceans of Enceladus may be teeming with life, and with water of a similar density and composition to our seas, maybe there are Enceladian whales, streamlined to cruise beneath the ice, as big as double-decker buses.

Titan, as far as we know, is the only place in the solar system apart from Earth that has flowing rivers. Lakes pockmark the surface, and the atmosphere is rich in carbon-based chemicals. While that all sounds rather Earthly, making Titan a good candidate for hosting life as we know it, those liquids are a mixture of ethane and methane, and the atmosphere is mostly nitrogen, with no oxygen but plenty of ethane, methane and acetylene, best known on Earth for welding metal. Basically, not a place for barbecues. Some have speculated that there might be liquid water beneath the petrolly surfaces, and that this might be a nursery for life. Liquid water is an essential ingredient for life as we know it, significantly because it can dissolve salts. Hydrocarbons such as ethane and methane are far less good as solvents. But yet again, these caveats arise from the limits of our imagination, combined with our knowledge of what life on Earth is like. With those energy-rich hydrocarbons sloshing around, on a dynamic planet there could be fish as big as whales, sucking up petrol to sustain their lives.

Like Enceladus, Europa has salty water under its icy crust. The oceans there contain twice as much water as our oceans on Earth, and possibly hydrothermal vents like those that dot the ocean floor on Earth, where they are the best candidates for the places where life actually started. Could those vents have worked like ours, so that, billions of years after the first spark of life on Europa, its oceans now also harbour beasts as big as their lunar constraints could enable, bigger than our blue whales?

Plans are afoot to send probes to Titan and Europa. NASA's *Dragonfly* is being designed to reach Saturn's moon, and may yet find Titans living there in its greasy atmosphere;

and *Clipper* is due to launch in 2025, its mission to explore the Jovian moon, and perhaps to discover what must surely be referred to as Europeans.

With those lunar liquids, there is no reason to suppose that the Enceladians, Titans or Europeans that swim around those moons might not be creatures akin to blue whales. Earth whales are a product of their environment. Their surroundings shape their size and form. We know this to be true – specifically for whales – because whales weren't always that big, and weren't always that wet.

An unfamiliar family tree

The creature pictured on p.50 is *Pakicetus*. Don't be fooled by its swishy long tail, whiskery snout, the furry milk teats hanging from its underbelly, and the fact that it's standing on dry land. *Pakicetus* was a whale.

Or at least, *Pakicetus* was the great-great-grandmama of whales. At some point around 50 million years ago, this creature – a sort of cross between a big dog and an angry otter, with four legs and a tail – started the process of getting back in the sea, 300 million or so years after animals first crawled out of the shallows and conquered the land.

Pakicetus was named after the place of its discovery: northern Pakistan, which today is known, among other things, for definitely not being underwater. At the time that our water-borne dog-whale was wandering the Earth, however, the geology of our globe was very different. This was before the then-island of what we know as India had very slowly crashed into the continent of Asia, a process which would

crinkle the Himalayas up to the sky. *Pakicetus* was loitering in
coastal waters; but its remains, relocated by the passage of
time and the shifting of the lands, are now buried a thousand
miles from the sea.

Why would a land animal become a sea creature, and thus
inadvertently give rise to a whole class of aquatic mammals?
It's difficult to know for sure, but perhaps *Pakicetus* found that
it was easier to escape from hungry predators by getting in the
water, or that there were shoals of delicious fish that it could
stalk in the shallows. Either way, the legacy of that land-based
past is still there in modern whales: they breathe air, they're
hairy (when born, anyway – they lose the hair quite quickly after
that), and they give birth to live young (as opposed to laying
eggs) which they feed with milk (which is more like a thick,
creamy paste than semi-skimmed, and is up to 50 per cent fat).

Once they got into the water, the descendants of the early
whales ballooned to almost 30 times the size of *Pakicetus*. The
salty ocean had made having a gargantuan frame possible, but
doesn't offer many clues as to why it might also be desirable.

The advantages of a whale's great size might be partly
to do with heat loss: the seas are cold, and water transfers

heat much more efficiently than air, so an aquatic beast will get colder more quickly than a land animal. As the descendants of *Pakicetus* evolved to become fully aquatic, they all developed thick blubber to prevent too much heat loss in the seas. The bigger the beast, the more slowly it loses heat, making the big, rounded whale extremely efficient at regulating its own temperature.

It's a persuasive argument, but surely heat loss can't be the only reason for the blue whale's size. Seals have blubber, too, and manage to thrive in the same aquatic environment while staying (comparatively) small.

One way we like to think about evolution is via a phrase not from one of the greats of the history of science, but from an American President. Theodore Roosevelt wasn't talking Darwin when he said, 'Do what you can, with what you've got, where you are,' but it very succinctly outlines the fact that evolution of organisms occurs in the environment, and can only do what is possible in that environment. The clues to how the whale got so big come not just from the working theories of heat loss, but also from where whales are and where they go – and the first clue comes not from a whale at all, but from a sly little hitchhiker.

The crustacean chronicles

Barnacles are born as little larvae that swim around trying to find somewhere to settle down. Once the baby barnacles are happy with the surface they've selected, they'll stick themselves on to it, using their foreheads, and build a crusty shell around them, staying put for the rest of their lives. Most of the time they'll find a permanent home on a rock on the seashore, but some barnacle themselves to the skin of whales and surf the oceans along with their giant steeds.[*]

Many modern whales travel huge distances with the seasons, gorging themselves on krill and a seafood buffet in the north Pacific during summer to build up lots of fatty energy reserves, and then heading south over thousands of miles to reach warmer waters for their winter holidays, taking their barnacle companions with them as they go.

During all those journeys the baby barnacles grow, drawing on the minerals in the surrounding water and slowly adding new layers to their hard outer shells. But sea water changes subtly depending on where you are – the profile of oxygen molecules in the south Pacific is distinct from those in the north – making the water itself an ocean fingerprint, absorbed by the barnacle into its shell. It means that the barnacle acts as a kind of passport, with stamps from all the places it has visited, allowing scientists to track the entire journey of a whale by carefully analysing the shells of its barnacle hitchhikers.

In 2019, a group of biologists had the idea of using this same technique on *fossilized* shellfish, sloughed off the bellies of whales millions of years ago. Looking inside the barnacle shells, they discovered the whales had travelled huge distances at roughly the same time as their bodies swelled in size.

It's just a working theory – we cannot go back and experimentally test whale evolution, alas – but the idea is that hoovering up tonnes of krill in the cold north during the

[*] Mating between two stationary barnacles some distance apart is understandably difficult. Nature always finds a way, however, and as a result barnacles have extraordinarily long penises, up to eight times their body length.

summer would provide fuel for the whales to migrate south into the warmer waters for the winter, and that is a lot easier if you are big enough to carry the necessary amount of fuel for such a vast journey. It looks, then, as if the size and shape of the king of the oceans were determined *by* the oceans.

Being big isn't easy

Size comes with a trade-off: big bodies might be able to eat more, travel further and retain heat better, but they also require a lot more fuel to keep them going. On land, meanwhile, where there's no buoyancy to help, and heat loss is far less of a problem, the downsides of being super-size are legion. No terrestrial animal has ever quite made it to the size of the blue whale, and for good reason. Only one ever came close – the real colossus of the dinosaurs: *Argentinosaurus huinculensis*.

Discovered in 1993 in Argentina, *Argentinosaurus* weighed in at a gargantuan 100 tonnes and stretched 40 metres long from end to end (roughly a Boeing 757 aeroplane/width of a football pitch) – a shade longer than the great blue whale, although that comparison is slightly unfair on the whale, since a lot of *Argentinosaurus'* extra length was winding neck and bendy tail.

Its great size made *Argentinosaurus* agonizingly slow – biophysicists estimate that at full tilt its maximum speed would've been around 5 miles per hour. That's about the pace of a stroll round the park. Although there are very few fossils to go on, it looks as if their necks were so long that they had to be lined with lung-sacs all the way down. This is a feature of giraffe necks too, without which they couldn't get enough fresh air to the lungs in the body cavity. As for supporting their vast weight, the fossilized remains show thigh bones that are 5 feet long, with a circumference of a staggering 4 feet. Being big isn't easy, and these are the kinds of physical adaptations big beasts need if they're to survive on land.

It may be that *Argentinosaurus* is as big as animals can get on Earth, given the constraints imposed by the planet's gravity. If there are beasts that big elsewhere in the universe, they'd have to be scaled accordingly, with colossal legs to carry all that weight – and they'd be utterly unlike the unimaginative lies of science fiction. You can't just scale up a giant version of a terrestrial insect, with spindly legs and a bulbous body. Super-sized spiders might terrify your dreams, but they're just not possible. If that's reassuring, we're not so sure: evolution on a different planet will have cooked up something much more frightening, even if it is much less familiar.

Spider-Man, Ant-Man and Man-Man

Reverence for the animal world is a good thing, and being impressed by the feats of beasts is perfectly justified. Sometimes, though, when writing about impressive animals, one can fall into the trap of false comparisons. Here's an example, taken from a popular animal website:

> **Tiny leafcutter ants can lift and carry in their jaws something 50 times their own body weight of about 500mg. That's the same as a human lifting a truck with its teeth.**

There's certainly no denying that the first half of this is true. Over the years, ants have been observed carrying leaves, dead

HOW STRONG IS AN ANT'S NECK?

How can you possibly measure the strength of an ant's neck? Clever measuring devices? Sophisticated calculations based on a deep understanding of biomechanics, ant anatomy and physics? Alas, no. Scientists worked this one out with an experiment, and we're sorry to say that it didn't turn out particularly well for the ant.

In 2014, researchers at Ohio State University took a number of Allegheny mound ants – a common American field ant – and glued their heads to the outside of a disc. Using a carefully calibrated centrifuge, they then spun the disc around and measured the force required to separate the ant's body from the head. 'That may sound kind of cruel,' the lead scientist acknowledged to an interviewer at the time, 'but', he was careful to add, reassuringly, 'we did anaesthetize them first.'

birds and all kinds of other insects and objects that are clearly many times their own body weight. Indeed, there's some evidence that ants might be even stronger – relative to body weight – than this statistic implies. The neck of an American field ant, for instance, can withstand forces of around five thousand times its weight (see box). The human neck, on the other hand, cannot.

Of course, we like to make comparisons between other animals and ourselves to help us understand what these animals can do, with statements like: 'If a flea were the size of a human, it'd be able to jump up to the 40th floor of the Empire State Building!' Even the most serious journals are not immune from this tendency: take, for example, this gem from the very august *Science* magazine:

A [dung beetle] can pull 1,141 times its own body weight – the equivalent of a 70-kilogramme person being able to lift 80 tonnes, the weight of six double-decker buses.

From such statements, it would be easy to get the idea that insects are the strongest creatures on Earth. But it's our job to disabuse you of any such perfectly reasonable but ultimately daft suppositions.

Sometimes in life, you can scale things up or down nice and neatly. If you spend twice as much of your pocket money on sweets, it makes sense that you'll end up with twice as many sweets. Walk double the distance up a street of identical houses, and you'll pass twice as many houses on the way. Triple the number of examples in a popular science book and you're three times closer to your contractual word count.

But, more often than not, scaling things up or down – as you have to do if you're making them bigger or smaller – doesn't mean you can calculate the resulting change in such simple arithmetical terms.

When it comes to changing animal sizes in particular, there's a good reason for this. Changing size means doing so in all three dimensions: height, width and depth. When it comes to strength, however, the most important factor is the cross-section of the muscle, which gets bigger only in two dimensions. When Dwayne Johnson pumps iron in the gym, his hunky biceps don't get longer – that would be ridiculous: they get wider and thicker. Strength doesn't scale in three dimensions, only in two.

All this means that if you want to scale an insect up to be a thousand times bigger in weight (that's ten times bigger in every direction: 10^3), it would likely only end up around a hundred times stronger (10^2).

Ants are strong not despite their size: they're strong *because* of their size. An ant weighing a puny 5 milligrams can carry a leaf of 250 milligrams; but scale it up to the size of a standard-issue adult male human, and it wouldn't have the strength to lift 15 kilos – far below its own body weight. Its legs would crumble as it tried to stand. In fact, these expanded ants would barely be able to lift their own heads and look us in the eye as they admitted our superior biology.

Sorry, Hollywood. This is why you can't simply scale up a gorilla to create King Kong, or scale up whatever Godzilla is meant to be to make Godzilla. It's why you won't find *Argentinosaurus*-sized mice or lobsters on an Earth-like alien planet. They'd all need to have significantly thicker limbs and fatter bodies to support their weight at such gigantic sizes.

How high can an alien jump?

Not everything changes with scale. Some things stay roughly the same, no matter how big or small a living creature gets – meaning there's a good chance that we'd find similar attributes in alien life forms. One example, perhaps surprisingly, is how high an animal can jump.

Assuming we're talking about a creature with some kind of legs here, then in the simplest terms, the amount of energy required to launch them into the air will increase as the creature gets heavier. But the amount of muscle mass the creature needs in order to generate enough energy to jump will also increase. These two components – the energy needed to launch and the energy available to launch – effectively balance one another out, leaving a fairly universal height that all creatures can jump to.

It sounds a bit counter-intuitive. You'd expect that humans should be able to jump much higher than insects – after all, we dwarf them by comparison – but in fact, the data on how high we can move our centre of mass from a standing start says otherwise, as you can see from the graph below. We're a little up on fleas, but about the same as locusts. Fleas, given that they're so tiny, are also affected by air resistance much more than the other creatures in the graph. Make a flea jump in a vacuum, and it'd be able to manage around 60 centimetres – just the same as us. It would also die.

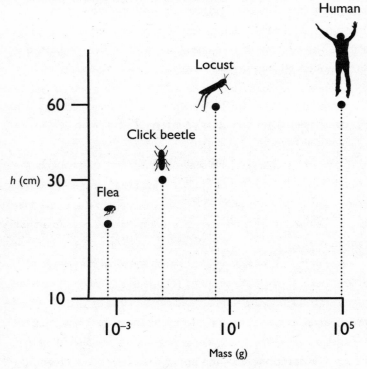

If a flea were the size of a human, it could jump as high as . . . a flea.

CAN ANT-MAN BREATHE?

These scaling laws work in the other direction too. This is another fact that appears to have bypassed Hollywood – and, in particular, the creators of one of the classic comic-book-on-the-silver-screen superheroes: Ant-Man.

The premise of Ant-Man is simple. A man, while in possession of a special suit,* gets shrunk to the size of an ant (too late for a spoiler alert?). While in miniature, our hero possesses superhuman strength and commands an army of ants.†

So far, so perfectly sensible. Scale a human down to the size of an ant and they would indeed be able to lift many times their body weight. Commanding an ant army via telepathy is a bit iffy, but we're willing to let that one go. But there's a problem, and here's where our‡ science pedantry overcomes our§ comic-book nerdery. Scale down a human lung and you don't just end up with proportionally less oxygen. The lung itself loses all of its efficiency and power. Lungs don't work at smaller sizes.

We're not alone in objecting to this particular oversight. This phenomenon was studied in detail in a scientific paper published in 2018, entitled 'Ant-Man and the Wasp: Microscale Respiration and Micro Fluidic Technology', in which the authors concluded that, because the volume of air we could physically inhale if we were shrunk down to a microscopic version of ourselves would be so small, Ant-Man and the Wasp should – in reality – be continually suffering from serious oxygen deprivation, roughly akin to being perpetually in the death zone of Everest. Funnily enough, the makers of the recent Hollywood film seemed to have bypassed the scenes in which Ant-Man suffers from headaches, dizziness, the build-up of fluid in the lungs and brain, and passes out.

*Adam – the comic-book fan among us – would like to point out that this isn't just any man. It's Dr Henry 'Hank' Pym, the original Ant-Man. Janet Pym (née van Dyne) also had a version of this power suit, but she had wings and so became the Wasp instead. Later, Pym would pass the suit on to reformed criminals Scott Lang and Eric O'Grady, but it's quite possible that you have stopped reading this footnote already.
† Hank Pym's technology also allows him to embiggen to become Giant-Man, but if he did that – just like King Kong – he'd have to have disproportionately thick legs. It's almost as if comic books are not science textbooks.
‡ Hannah's.
§ Adam's.

Dead, broken or splashy

Size isn't just another attribute, like colour or texture. It isn't something you can play around with and tweak at will. Things are the dimensions they are because that's the size they need to be.

That's why you won't find any mice or lizards in Arctic regions, and why polar bears and walruses can thrive in the cold. Shrink a creature down, and its surface-area-to-volume ratio will leave it unable to regulate its own body temperature.

It's also why insects are less affected by gravity. Smaller creatures have more surface area relative to their mass, so if you're small enough, your body can effectively act like its own parachute. You could, in theory (but please don't), chuck a mouse out of a plane and – as long as its landing was soft – see

PERSONAL PLUMBING

There's another, more intriguing example of something that doesn't change with size – something known as 'The Universal Law of Urination'.

In 2014 a group of scientists published a paper in which they tackled a question that literally no one else had ever asked. They set out to 'elucidate the hydrodynamics of urination across five orders of magnitude in animal mass'. Or, to put it another way: they watched loads of videos of animals weeing and timed how long it took.

The team found that all mammals take roughly the same time to empty their bladders, regardless of how large they are. Big animals, like elephants, have bigger volumes of wee to get rid of – but they also have longer urethras and faster flow speeds, being subject to greater gravitational forces (imagine a firehose and you're pretty much there). Much smaller animals, like mice and bats, have to battle against viscosity and surface tension in their urine, meaning they can only excrete a single drop at a time. These competing factors balance each other out, to leave a near-universal time to empty a full bladder of 21 seconds.* Try it yourself next time you nip to the loo.

*Error bars: +/– 13 seconds.

it escape pretty much unharmed. Indeed, the twentieth-century biologist J. B. S. Haldane considered this very question in a thought experiment involving not a plane, but chucking animals down a well shaft. (We assume – we *hope* – it was a thought experiment.) He concludes that a mouse would walk away, but a 'rat is killed. A man, broken. A horse *splashes*.'

All this is also why giraffes and sauropods need lung-sacs lining their necks. An insect can get away with absorbing oxygen by osmosis through its body surface, but for every tenfold increase in overall size, you need a thousand times more oxygen to fuel the cells, with only a hundred times more surface area through which to suck it in. Lungs and gills are nature's way of adding oxygen-absorbing surface area. The human body has around 180 square metres of lung squeezed into the chest. Scale up to the size of a diplodocus, and you've got a creature that's yearning for every spare millimetre of lung just to allow it to breathe. In Haldane's own words: 'The higher animals are not larger than the lower because they are more complicated. They are more complicated because they are larger.'

That's the problem with getting too big: physics has non-negotiable rules. Evolution does its best to invent useful workarounds. But there comes a point where all the circulatory systems that keep creatures upright – blood flow, oxygen flow, nerve impulses – start to become infeasible. Together, they leave scientists in little doubt that *Argentinosaurus* and the blue whale are close to as big as it gets on Earth. There is a natural upper limit to how big a beast can be before gravity wins – and in the end, gravity always wins.

On a smaller or lighter planet, where gravity was weaker, a less forceful force could throw Haldane's ideas out of the window. Without the weight of a body bearing down, legs could be much thinner. Insects on a non-Earth-like planet could be much taller with spindly long legs, cows more like giraffes, and giraffes more like something out of a Salvador Dalí painting. Trees could tower like skyscrapers, hundreds of metres tall.

Those trees would have to be dramatically different from Earthly plants, however. The weaker gravity would mean that the soil wouldn't drain efficiently, and roots would easily become waterlogged. Indeed, this is precisely what green-fingered astronauts in the International Space Station discovered when they tried to grow Chinese lettuce, zinnia flowers and a whole (small) garden in the microgravity of low orbit.

Doing anything in microgravity is a tricky business, but growing plants is particularly hard. Earth-plants have evolved over the last 2 billion years to grow in soil, and soil is a bit crumbly and will therefore float around in your space capsule. Having loose matter floating around in your space capsule is a really bad idea. So instead the seeds are planted in a gel, a bit like the padding in a disposable nappy. The gel stays moist, without saturating the roots of the plant as wet soil would.

So perhaps alien flora sit in compost with all the properties of the inside of a pair of Pampers (though this seems unlikely). Or perhaps the liquid that seeps through the alien vines would have to be actively pumped by the plant itself. Anything is possible.

To test the robustness of plant life on alien worlds, the Chinese lunar explorer Chang'e 4 contained a mini-menagerie, which landed near the south pole of the Moon on 3 January 2019. Inside the 18-centimetre cylinder were seeds for potatoes, some samples of *Arabidopsis* – a type of cress beloved of plant scientists – and rapeseed, which we grow for oil here on Earth. The plan was not to make a potato salad, though. There were some fruit-fly eggs and some yeast in there too, and the idea was to see if a mini-ecosystem could flourish, the flies breathing out CO_2 that would nourish the plants, which in turn would produce oxygen, and the yeast helping to regulate the

VICTORIAN GARDENERS IN SPACE

The International Space Station has been occupied by humans since its construction in 1998, a satellite home for more than 230 women and men looking down on us from above. But the first description of a human-made satellite comes from fiction. The 1869 sci-fi novella *The Brick Moon* by Edward Everett Hale featured a 200-foot-diameter hollow sphere made of bricks, to be launched by rolling it down a giant slope into a pair of flywheels that slung it into a low Earth orbit. Its function was to act as a navigation beacon fixed in space, like the North Star does for sailors. Alas, it slipped from its foundations and launched itself by accident ahead of schedule, and thus took with it the 40 families who were living inside the brick orb during its construction.

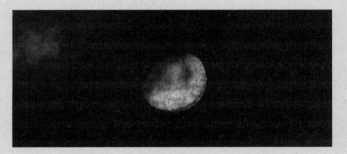

But, just like Mark Watney in *The Martian*, the accidental astronauts survive and flourish by cultivating soil and growing plants; they feast, party and communicate in Morse code by hopping around between giant lunar palm trees, and walk from summer to winter whenever they fancy a change of season. Frankly, it sounds like heaven.

atmosphere. We await the conclusion of this experiment, which may not be revealed until a human returns to the Moon in the next few years and cracks open this mini-biodome. We predict that one of the three following possibilities will have come about: (1) everything is alive and thriving, and this will mark the beginning of a new era of space botany; (2) everything is dead, killed by space radiation and the weirdness of a gravity that is alien to the organisms within, leaving a mess that will smell like bin juice; (3) the organisms have cross-fertilized in some inconceivable science-fiction way, creating a mutant fly-potato hybrid hell-bent on destroying humankind. Place your bets.

We want to believe

And so we circle round to where we began. It's fun to speculate about life off Earth, but we are hideously restricted by two things: (1) life on Earth is bounteous, bonkers and wonderful, but it is the only life we know of; and (2) we are not nearly as imaginative as evolution. We find it very hard to think beyond the only life we know of, especially when aliens are such a beguiling idea that our fiction is littered with them. Is there life in the rest of the universe? We don't know. If there is, it may simply be too far away for us to ever discover it. If there isn't, that makes the Earth an even more precious place, and we should redouble our efforts to protect it.

We like to think that the universe is teeming with life; otherwise it seems like an awful waste of space. There are some things we can say with reasonable confidence, though – this is scientific speculation, not guesswork, and we're happy with that.

Most life is small – being big is hard. If extraterrestrial life forms are big (that is, bigger than bacteria), then they will probably have photoreception. Being able to see in some capacity is enormously helpful in finding stuff to eat, and avoiding being eaten. If you can see, then colour opens up as a range of possibilities, so life will be colourful. They will have guts – a way of internalizing food to efficiently extract all of its nutrients. If it's cold, they'll be round. Ish. If they're based in liquids, they're likely to be torpedo-shaped, with a tail or some other form of propulsion. When *Pakicetus* was dipping her toes in the shallows, sharks looked pretty much like they do today. But getting back into the oceans streamlines whales to look a lot like sharks, because that's what works best in the seas. If they fly, they will have wings, just like birds, pterosaurs and bats do, even though they are hundreds of millions of years apart. If they're on land, they will definitely have legs. Maybe six. Or more. Or none.

Or it might be none of the above, but they'll definitely wee for 21 seconds.

Life evolves in its environment. An organism is crafted by the cosmic happenstance of where it lives. This is as true for us as it is for every living thing.

If we were to run human evolution all over again, and imagine that our nursery was not the forests and plains of Africa but jagged rocks, or swamps with 100-foot vines, would we have evolved to have better climbing feet, or even retained our tails? In the swamps, would we have developed splayed feet and hollow bones to wade in shallow waters and aid buoyancy? These questions are unanswerable. We are what we are, because of the climate, the landscape and the planet where it all happened.

Those three words – life on Earth – are so powerful, and so evocative, reminding us that we are just a tiny branch on a ridiculous family tree that spans 4 billion years, six great extinctions and more creatures than we can imagine or count.

But the 'on Earth' bit can be easily overlooked. It's not just that life exists *on* Earth. The Earth shaped life. Life is the way it is *because* of the Earth, because of its size, its dimensions, its gravity, its distance from the Sun. We toy about with ideas of aliens because it's fun, and because there's a big question about life elsewhere in the universe. But really, we think about extraterrestrial life because it tells us about us, about evolution on our precious space-rock. Not 'life on Earth' but 'life *and* Earth'.

CHAPTER 3

THE PERFECT CIRCLE

Fritz Zwicky is famous for two things: first, his pioneering cosmological work in the 1930s studying gravity, the Big Bang, dark matter and neutron stars; and second, for being a curmudgeonly git. Alongside all the cosmos-quaking physics he did, Zwicky coined the insult 'spherical bastard' for colleagues he didn't like, on account of their being bastards from whichever direction you looked at them.

Spheres come in all sizes, but not in all shapes. However you twist and turn them, they are all, in essence, identical. Look at them from whatever direction you like; they are always the same. It's why the ratio between their diameter and circumference is always the same, too – this ratio being where the number π came from. This may seem very obvious, but one of your authors (can you guess which?) only worked this out as an adult.

The universe is full of spheres: planets, bubbles, footballs, annoying colleagues. Roundness abounds. And yet, just to remind you of the theme of this book, much of what you might think you know is not quite what it seems. Spheres and circles are more of a mathematical fantasy than you might expect; and to see why, we're going to have to go on a quest through everything on Earth and beyond, from the structure of the atom to the fabric of spacetime itself, stopping off via Newton, Einstein, a space squid and a suspiciously large nipple.

What does a sphere look like in four dimensions?

Before we get to all that, though, here is some mind-bendy weirdness, essential for considering the sphere, in the form of four-dimensional balls.

Your instinct might be to try to picture what a 4D ball looks like: resist that urge. The first rule of imagining what something looks like in four dimensions is to stop trying to imagine what it looks like in four dimensions. We're completely constrained by our inescapable three-dimensional reality; so however hard you try to force your mind's eye to see something beyond everything you've ever known, you're only going to end up frustrated, and probably quite confused.

Even if you can't make an image of it in your head, though, it is still possible to go a long way in describing what a four-dimensional shape looks like. The trick is to pay careful attention to what happens when you shift from two to three dimensions, and then apply the same rules when you move from three to four.

Start with the simple: a two-dimensional sphere is, of course, a circle. Indeed, mathematicians often don't call them circles at all, but 1-spheres. (What you and I would just call a normal sphere is a 2-sphere.) You can easily get from a circle to a sphere and back again: a spherical ball can be made by stacking lots of circles on top of one another in steadily increasing and then decreasing sizes (a bit like making a ball out of Lego bricks, or a Pac-Man out of pixels). A circle can be created by taking a very thin slice through a ball.

This is the key idea you need to imagine the next dimension up: that circles are two-dimensional slices through three-dimensional balls.

3D printers work on precisely this notion of slices, building thin layers one on top of the other, the printing arm moving slowly up through the machine as each new layer is added.

Imagine, then, that you are a tiny two-dimensional creature – like a very flat ant – sitting inside the 3D printer, on a special flat-ant-sized platform on the printing arm. You can't see up, or down (you're too flat). All you can see is the new layers of the object being added one by one by the printer, each thin enough to be practically two-dimensional.

If the 3D printer were building a ball from scratch, you (that's ant-you) would witness the arrival of each new layer as a circle.* The circles would start off small as the printer built the bottom of the ball, getting larger towards the middle and then

---

* Technically, if the ant were fixed in position, it would only see a line curving away from it. A two-dimensional object can't see the whole two-dimensional circle at once if it's sitting in the same plane. It's the same story for us in three dimensions. When you look at a ball, are you really seeing a three-dimensional sphere? Or are you seeing a circle that curves away from you? One that you merely know from experience is a sphere?

small again near the top. As each new circle is printed, the arm moves up to build the next layer, taking you with it – so that the previous circle vanishes from your view.

That's the story in going from two dimensions to three – circles of slowly changing sizes appearing and disappearing in front of your little ant-eyes. There is, therefore, no logical reason why the story would be any different in going from three dimensions to four. It's just a matter of bumping everything up by one dimension. Which means the following strange idea is true: spheres must be three-dimensional slices through four-dimensional balls.

Now imagine you are a three-dimensional being (that shouldn't be too taxing). The next bit is slightly harder. Imagine you're standing inside a 4D printer as a 4D ball is being made. What would that process look like?

First, you wouldn't be able to see the whole printer – only a 3D slice of it. Just as the ant had no way to look up or down, there's an extra unimaginable dimension (this time, without a name) that you wouldn't have access to, or even be aware of.

As the printer whirred into action, the first thing you'd see is a very small, but very perfect ball being printed; then, immediately it's finished, the ball would vanish into thin air. It has disappeared into another slice of the fourth dimension. There's no time to wonder where it (or your sanity) went, because the printer is already on to layer two, producing another perfect ball, this one slightly bigger than the last. It's there for a moment, and then pouf! It's gone.

Layer after layer, the balls get bigger and bigger, each in turn disappearing as soon as it is made, until you reach the centre of the 4D sphere, at which point the printed balls start getting smaller again. All these successive layers (which are actually, in themselves, 3D balls) are sticking together in a dimension you can't see, to build a shape that your puny 3D mind isn't capable of imagining.

There are still sensible things to say about the 4D ball that's just been printed, even if you can't see it. We could plot it on a graph. Rather than having a centre at (0,0) as a circle would, or (0,0,0) as a ball would, our 4D sphere has a centre at (0,0,0,0).

Or consider its shadow. If you hold a torch above a physical object, the shadow it makes on the ground is a flat projection of the object. To put that another way, the shadow cast by a ball is a circle. Cast a shadow and you see the same object, projected into a lower dimension.

That means – hold on to your brains here – that the shadow cast by a 3-sphere would be an actual ball. Our 4D printed sphere, if held above our world in an extra dimension, would cast a perfect three-dimensional, ball-shaped shadow, like a floating orb of darkness, a perfect sphere of shade.

Such a thing sounds like nonsense, but reality is sometimes stranger than fiction. Much of physics plays with the idea that we are not living in a three-dimensional world at all, but in one with up to 26 dimensions – a universe where hyperdimensional spheres and orb-like shadows are everywhere.

Is anything perfectly round?

Zwicky's insult – 'You spherical bastard' – works because only things that are perfectly round can be exactly the same from every possible angle. Most round things are near enough to do the job they're designed to do – wheels on the bus, records spinning on a turntable, footballs (though in 2009, Nike did claim to have designed a ball that was actually rounder than others). But we like precision, and all those things are actually only round-ish.

In the hunt for circular perfection, there are some strong contenders: raindrops, bubbles, water ripples and rainbows – these are a few of our favourite things. Each of them comes close, but upon inspection they all fall short. Raindrops and bubbles both adopt approximately spherical shapes thanks to surface tension – the forces within seek to find the lowest-energy state possible by smoothing out lumps and bumps and pulling in any corners. But in reality, both are buffeted by wind and distorted by gravity, so will never truly adopt spherical perfection. Water ripples, too, are subject to the distortions of the unruly environment. Rainbows might seem to have a better claim – particularly that rarest of beasts, the circular rainbow, which you might have been lucky enough to spot from the window of a plane if the conditions were just right. A rainbow, though, isn't really a 'thing'. Rather, it's an illusion created by the arrangement of a vast number of water molecules refracting the light of the Sun.

How about the world of the living? In nature, making yourself very round might seem like a good idea – as we just said, it's the lowest-energy state possible, and since the act of living requires energy, the less of it you need, the easier it is to stay alive. Hedgehogs, armadillos and woodlice (known in the US by the adorably cute name of pill-bugs) all scrunch up into a ball when they're threatened – thus minimizing the surface area on their bodies that can be attacked. So maybe biology, with its unending quest to find the easiest way to continue living, is the best place to look for a perfect sphere.

Many of us get our first hands-on experience of biological circles at school, when we're given a microscope and invited to peer at a leaf, specifically to see if we can spot the round holes on the underside through which the leaf breathes. Like a lot of things in biology, these little holes have been given a Greek or Latin name, partly to aid the scientific description, but probably mostly because Greek or Latin sounds a lot cleverer than English. Exhibit A: these holes are called stomata, which is the Greek word for mouths.* At first glance, stomata look fairly round, but closer inspection reveals they aren't: they open and close, like mouths, and are more oval than circular.

As for cells, in humans the egg is the biggest and probably the most spherical,† although it is nigh on impossible to determine how spherical, because we can only look at these gelatinous bags in two dimensions down a microscope: they are too small for us to hold one up in three dimensions and assess how spherical it is. In truth, eggs don't even come close.

Even if an egg starts out more or less spherical, it doesn't stay that way for long. On its route to becoming a person, it's invaded by the successful sperm, headbutting its way in, past the electric fence that surrounds the egg. In mice – we don't know whether this applies to humans yet – the point of entry of a sperm into an egg determines which cells will be placenta and which will become a baby mouse, and which end will be head and which will be tail.

If cells aren't spherical, what about organs? The eye, for

---

* If you still need persuading, let us present Exhibit B: the socket at the base of your skull, through which your spine is attached to your skull. It's called the *foramen magnum*. Translation: big hole. The prosecution rests its case.
† Sperm also take home a superlative, winning the trophy for the smallest human cell.

example, which looks at first like a good candidate, is let down by the anterior segment – that's the bit that sits on the front with the cornea, iris and lens, helping us to focus. The shape of the eye, it seems, also depends on what sort of environment the owner is in. In space (for reasons we don't yet understand) eyeballs get distorted even further away from a spherical form. All astronauts come back to Earth needing glasses for at least a few months because their eyes have become elongated, meaning that the focal point generated by the lens is in front of the retina, which causes short-sightedness. Most astronauts recover to 20:20 vision, but some don't. Doug Wheelock, who spent a total of 178 days in space aboard the Space Shuttle *Discovery*, the International Space Station and the *Soyuz* spaceship, was grounded by NASA after his vision failed to recover, and now wears glasses permanently. His nickname, by the way, is Wheels.

If you're looking for circular perfection in living creatures, it's a non-starter. There are too many peculiarities, and too many subtle complexities, for biology to ever quite round up roundness.

How do we know the Earth isn't flat?

Let us leave Earth-bound life forms and instead consider the planet itself: is there a terrestrial-sized contender for the perfect sphere?

It wasn't until 1522 that Ferdinand Magellan circumnavigated the planet, 30 years after Christopher Columbus had made it to the Americas. We have, however, known the Earth is round for more than 2,000 years.

THE MATHEMATICIAN WHO TURNED DOWN A MILLION

At the turn of the millennium, the Clay Mathematics Institute in America set a challenge to the rest of the world. It published a list of the hardest and most important unsolved problems within mathematics. The list contained seven questions, and anyone who could solve just one of them would be offered the prize of $1 million.

Cast aside any images of the tricky problems you tackled at school, or even university. These are questions of boss-level difficulty. For instance, one of your authors spent her entire PhD studying the exact same equation that is the subject of one of the Clay millennium problems, and – to this day – she only just about understands the question. They are, to put it mildly, ridiculously hard.

So hard, in fact, that only one of them has been solved as yet, and it was a problem all about spheres in four dimensions.

It's known as the Poincaré Conjecture, and comes from a branch of mathematics called geometric topology, part of which essentially involves mathematicians imagining what it would be like if everything were made from plasticine. There are rules in this plasticine world: you can mash and bend objects as much as you like, but you can't add or take away holes. The question is, in such a world, which objects are similar to which others?

In this way of thinking, a cube is the same as a pyramid, because you can easily squish one into the other (if it's made of plasticine). The corners and edges don't matter – you can flatten, reshape and rebuild them at will. Though it's slightly harder to imagine, a coffee mug is the same as a doughnut (both have only one hole – through the middle of the doughnut and the handle of the mug – so the cup part can be flattened and subsumed by the handle to become the doughnut). Harder still to visualize, but nonetheless true, a T-shirt is the same as a doughnut with three holes. You can pull the bottom of the T-shirt, stretch it and stitch the edge to a hula hoop to reveal three holes inside – a bit like a fidget spinner.

If you play this game, you rather quickly realize that a solid 3D object without any holes could – if you wanted – always be sculpted into the shape of a ball. Poincaré's Conjecture asks: is the same true in four dimensions?

This was a question that stood unanswered for almost a hundred years until, in 2003, a strange proof appeared on the internet, posted by a little-known Russian mathematician called Grigori Perelman. Early on, many serious mathmos were dismissive of Perelman's solution – there are lots of people on the internet who have claimed to have solved the Conjecture, and most of their 'proofs' consisted of a few garbled pages of nonsensical gibberish. But slowly, more and more people began to take an interest in Perelman, and momentum began to build as people realized his proof was quite possibly the real thing.

It took three years to check his working carefully, but once the confirmation came through in 2006, Perelman was offered the $1 million.

Which he immediately turned down.

The maths community also tried awarding him the Fields Medal (often described as the Nobel Prize of mathematics, but much harder to win – it's only given out every four years, and even then only to someone under 40). Grigori Perelman had won the most prestigious prize in all of mathematics, and again he turned it down.

Perelman said he had no interest in being recognized by the mathematical community. He didn't want to be treated like an 'animal in a zoo', as he put it; to be stared at and have his life pored over by mathematicians less accomplished than he was. Rumour had it that the location of the Fields Medal ceremony was also something of a problem. The prize was due to be awarded in Madrid. That would mean Perelman taking a day out to travel from his home town, Moscow, a day for the ceremony and a day to return home – three days when he could have been doing maths instead. Grigori Perelman: we salute your commitment.

Eratosthenes was born in the third century BCE in what is now Libya, then part of Greece, and so, growing up, he was tutored in intellectual pursuits in the local school, where they also did youthful things such as naked wrestling and probably discus-throwing. He then went off to Athens, where he studied Plato and wrote well-regarded poetry, drew up a chronicle of the Trojan Wars and compiled a chronology of the winners of the Olympic Games – effectively the first sports almanac.

In a classic case of ancient Greek banter, some of his contemporaries called Eratosthenes 'Beta' – the second letter of the Greek alphabet – because they thought he was second-rate compared to other thinkers of the age. Strangely, no one remembers those guys.[*]

As his career progressed, Eratosthenes was eventually selected for the extremely prestigious job of head librarian in Alexandria, Egypt, the great scholarly heart of the Mediterranean, and it was here that he did his most enduring scientific work, including the first estimate of the size of the Earth using only a sundial and a bit of maths.

While in Alexandria, Eratosthenes heard a story from travellers about a well in the town of Syene (modern-day Aswan in southern Egypt). At midday on the summer solstice – the point in the year when the Sun is highest in the sky – the light from the Sun would hit the water at the bottom of this well spot-on, without casting a shadow. Indeed, none of the buildings or rocks or any other objects in Syene had shadows at that time on the solstice. At that moment, therefore, the Sun must be sitting perfectly overhead.

Most of us would hear that story, think it was interesting, then get on with our day. But it gave Beta an idea: a brilliant experiment based on the path of shadows. First, he wondered whether shadows appeared in Alexandria at noon on the solstice, and so tested this by sticking a gnomon – basically, a vertical rod – in the ground. He noticed that, unlike anything in Syene, it was throwing some shade.

Eratosthenes figured that if the Earth was flat, at midday on the summer solstice no shadow would be cast at either location. And yet here it was: a shadow at Alexandria, at the precise moment there was no shadow in Syene to the south.

[*] See also Plato. His real name was Aristocles. 'Platon' means broad or wide, and was given to him as a nickname, probably because he was wasn't thin or narrow.

The only possible explanation was that the surface of the Earth was not flat, but curved.

In Alexandria, Eratosthenes' gnomon cast a 7-degree shadow – just under one-fiftieth of a circle. He then gave some poor soul one of the worst jobs in scientific history – paying him (at least, we hope he was paid) to pace out precisely the distance from Alexandria to Syene, which came out as 5,000 *stadia*, roughly 800 kilometres or 500 miles.*

It was then a simple calculation: if the distance between Alexandria and Syene represented one-fiftieth of a circle, then the circumference of the Earth would be 50 × 5,000 *stadia*, a total of 250,000 *stadia*, approximately 40,000 kilometres. The actual circumference of the Earth as measured today around the equator is 40,075 kilometres.

That was impressive; but it was bettered by the work of the Iranian scholar Al-Biruni. Born in 973 CE and one of history's greatest scientists, he came up with a startlingly accurate estimate of the circumference of the Earth, without needing to rely on the error-prone method of measuring a long distance, like that between Alexandria and Syene.

Al-Biruni's idea was first to work out the height of a mountain. For this, he used an astrolabe – a disc with a rotatable arm, not totally dissimilar to a protractor, but made of brass and much more ornate. Standing at the base of the mountain, he used the astrolabe to measure the angle to the summit.

* There is some dispute about how long a Greek *stade* was, though it was measured every year by pharaonic book-keepers, possibly by timing camels on a specific journey. The distance does change over time, which also changes the error margins of Eratosthenes' calculation, leaving him between 10 per cent and 15 per cent out. Still, not bad for an old Greek dude.

He then walked a short distance from the mountain, 100 metres or so, carefully recording the distance along the way, and repeated his astrolabe measurement from that point.

From those two measurements, Al-Biruni had enough information to calculate the height of the mountain. (For those interested, see if you can work out the trigonometry he used yourself. The answer is in the next box.)

Once he had this measurement, Al-Biruni only needed one more number to give him the radius of the Earth. Carrying his trusty astrolabe, he climbed the mountain and, from the summit, measured the angle between the horizontal and the horizon, and thus had the final ingredient for a gigantic planet-sized right-angled triangle, based on a single unknown factor: the radius of the Earth.

With just basic trigonometry, Al-Biruni calculated that the radius of the Earth was 3,928.77 miles. Today, the best calculation of the mean radius of the Earth is 3,958.8 miles.

So, whether we look back to ancient Greece or the Islamic Golden Age, we've known our planet is not a flat disc for a very long time, despite what a troublingly numerous – and growing – club of conspiracy theorists might say.

Eratosthenes' and Al-Biruni's brilliantly simple experiments showed beyond doubt that the surface of the Earth is curved; and on this basis they assumed – perfectly reasonably – that the Earth was perfectly spherical.

But that question – whether our planet is genuinely spherical rather than just round-*ish* – has provoked a surprisingly large amount of debate over the years, with some serious practical consequences.

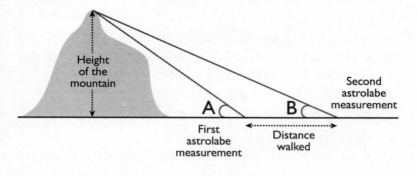

Height of the mountain

A

First astrolabe measurement

Distance walked

B

Second astrolabe measurement

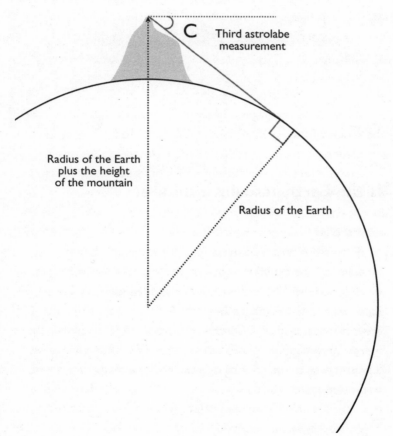

C

Third astrolabe measurement

Radius of the Earth plus the height of the mountain

Radius of the Earth

THE GENIUS OF AL-BIRUNI

OK, maths fans, time for some answers.

If *A* and *B* are the first and second angles measured by the astrolabe respectively, and *D* is the distance walked by Al-Biruni, then the height of the mountain, *H*, can be calculated.

$$H = \frac{D \tan A \tan B}{\tan A - \tan B}$$

And once you know *H*, and the third astrolabe measurement C, the radius of the Earth, *R*, follows on nicely.

$$R = \frac{H \cos C}{1 - \cos C}$$

Is the Earth actually a sphere?

Christopher Columbus wasn't the kind of man who doubted himself. One of history's most arrogant and brutal tyrants, he believed he'd been personally chosen by God to cross the Atlantic and sail around the globe, because no one else on Earth was up to the job. Setting off west from Spain, he and his crew made landfall on a Caribbean island in 1492, becoming the first colonizers of what they called the New World. Of course, it wasn't new to the millions of people who already lived there, who were systematically conquered and slaughtered, or died from diseases their immune systems had never encountered. Their possessions were plundered by the invaders from

Europe, and anyone who objected was subjected to one of Columbus's favourite tactics against dissenters – which was to cut off their hands, hang them around the offender's neck and send them back to their village as a message to discourage rebellion. Columbus was a monster.

Given that humility was not one of his stronger character traits, it was all the more remarkable that in 1498 he started to wonder if he had made a major mistake.

It's worth noting, while we're on the subject, that while Columbus and his men were the first Europeans to *permanently* colonize the Americas, they weren't the first to get there, by a long way. Around the year 1001 CE, the Icelandic Viking Leif Ericson landed and set up camp in places that he and his companions called Markland, Helluland and Vinland – which we now presume to be Labrador, Newfoundland and Baffin Island in Canada. The Americas had been peopled probably for 20,000 years before the Vikings showed up, and these lands were already occupied by indigenous people the Vikings called the Skraeling, who we think were probably the ancestors of the Inuit. Thorfinn Karlsefni and Gudrid Thorbjarnardóttir were among Ericson's gang, and around the year 1004 CE they had a son, Snorri Thorfinnsson, who was the first person of European descent to be born in the Americas. But their holiday didn't last. Leif and his people stuck it out for three years before they decided that the Skraeling were too unruly for them, and, following an argument about a rampaging bull, they upped sticks and left. The next time a European set foot on that continent was in 1492.

Back to Columbus. He was familiar with Eratosthenes' proof, and used it as part of his planning to find these new lands to conquer and plunder. He too was convinced that the Earth

was round, but he had elected to use a different calculation
– one of his own making – and that gave him a figure for the
size of the planet about 25 per cent smaller than Eratosthenes'
estimate. It was because of this calculation that he confidently
believed his ships had reached the Far East, when in reality he'd
barely gone past Cuba.

As he neared his 50th birthday, Columbus found himself
sailing around an island that he thought was in Asia, but was in
fact just off the coast of Venezuela, when he started to notice
something peculiar.

The weather was strangely clement, which didn't make
sense for a position so close to the equator. When he checked
his position by referring to the North Star, Polaris, it seemed
to be moving erratically in the sky, and there was a veritable
river of fresh water flowing into the sea around his ship. It
didn't feel like he was navigating around the globe. It felt like
he was sailing *up* it.

By now (in fifteenth-century terms), Columbus was a fairly
old man. His fame and fortune were founded on the round
Earth theory. But, here on the ship, he was confused by the
evidence he was collecting, and with this he began to question
everything he thought he knew. In a letter to the King of Spain,
he explained:

**I have always read that the world comprising the
land and water was spherical, and the recorded
experiences of Ptolemy and all the others have proved
this. But I have now seen so much irregularity that I
have come to another conclusion respecting the Earth,
namely, that it is not round, as they describe, but of
the form of a pear, which is very round except where**

the stalk grows, at which part it is most prominent; or like a round ball, upon part of which is a prominence like a woman's nipple.

That nipple part, incidentally, he believed would be closest to heaven, hence the temperate climate.

Columbus's nipple-Earth theory didn't stand up to scrutiny. But behind its absurdity, and if you're willing to ignore the detail, there is something admirably scientific about his coming to such a conclusion. Columbus was far from being an honourable or likeable man; but, on this point at least, he showed himself willing to dismiss his own life's work on the basis of new evidence – relying not on what he'd seen, but on what he had measured and calculated. Even though in fact the Earth is not pear-shaped, and has no prominent nipples.

It's not a sphere either. The hills and valleys, mountains and ocean trenches on the Earth's surface make our planet too lumpy and bumpy to be a perfect sphere – although they make less of a difference than you might imagine. Even something as big as Mount Everest is only a minor blemish when you consider just how big the world is as a whole. Indeed, if the Earth were shrunk down to the size of something that could fit into your hand, it would be considerably smoother than a billiard ball.

But it wouldn't be rounder. There is something going on that's much more dramatic than the odd mountainous blip here and there. If we take a large interplanetary step back and have a look at the Earth from the distance of space, we see that our globe is not even close to being round. Our planet is a bit porky round the middle and a bit flat at the top and bottom. It's not a sphere at all, but an oblate spheroid.

And what is the reason for this middle-age spread? It's that the Earth is an active, rotating planet. Isaac Newton thought that because the Earth spins, and is fluid under the mantle, this would force it to bulge at the equator. He was quite right: according to the latest measurements, the distance from the centre of the Earth to sea level is about 13 miles longer at the equator than at the poles.

Twenty-first-century techniques also show us that the Earth isn't even perfectly oblately spheroid either. Scientists fire lasers at satellites and check the rebound times, and listen to radio waves from outside the Milky Way to see how they change as the Earth moves and reshapes itself. The gravitational pull of the Moon and Sun cause oceanic and atmospheric tides. Plate tectonics means that the mass of the Earth is not distributed evenly. There's even a phenomenon called 'post-glacial rebound', whereby the mantle springs back (at a rate of a few centimetres every thousand years or so). Just as the planet is pulsating with life, it's throbbing in itself.

Spheres in space

So this blue marble floating in space that we call home is delightfully knobbled. Our nearest neighbour, Mars, is not as oblately spheroid as Earth. It was once a thriving volcanic planet, and is now dead, but its tumultuous history has left it with much bigger pimples on its face than we have on ours. Olympus Mons is an extinct shield volcano, which means that the lava it ejected didn't spurt into the air but dribbled out over millennia, layering the sides of a growing mountain. As a result of all this gradual ooze, Olympus Mons rises at a gentle slope, about a 5 per cent incline – so one foot up for every 20 feet forwards, which is more like a hike than climbing Mount Everest. That is not what makes it a big deal, though. Everest – the tallest mountain on Earth – is a mere 5.5 miles above sea level, and very steep. Mauna Loa, Earth's mightiest volcano, is 6.3 miles tall, and only 2.6 miles of that is above sea level. Olympus Mons dwarfs these pimples: it is 16 miles high from its base.

Near the top, it has six calderas – the collapsed craters of what were once lava blow-holes – and its area at the base is about 120,000 square miles, meaning that it would just about fit inside the borders of France. The sheer size of this sleeping monster zit means that on its own it distorts the planet enough to prevent it from being a sphere.

Other planets are similarly wonky. The gas giants Saturn and Jupiter are even more oblately spheroidal than the Earth, meaning that they are more squashed at the top and bottom. Saturn's polar diameter is only 90 per cent of its equatorial diameter – if it were a bowling ball it would wobble hugely, and certainly not roll in a straight line. Even its iconic rings are not perfect circles but elliptical and changing, because the gravity of moons such as Titan squashes their sides.

These two gas giants are less spherical than the Earth because they are gassy and giant. The biggest gaseous ball in our local neighbourhood is the Sun, our star, and it is probably also the most spherical natural object we have measured. A definitive study, published in 2012, which was set up specifically to find out the shape of the Sun, concluded that its shape does not change significantly during its own internal 11-year cycle of activity, and that the difference between its diameter at the equator and at the poles is only

6 miles out of 890,000. The apparent clarity of this conclusion, though, is slightly blurred by the fact that we aren't quite agreed on where the Sun's boundaries are. Solar flares many times the size of the Earth are spat out into space often but unpredictably. What we can see is only a fraction of the Sun's reach, which envelops Mercury and Venus, depending on what filter you have on your camera, and envelops the Earth as well if we're talking magnetic fields.

Nevertheless, despite its invisible edges, we can legitimately say that the roundest naturally occurring thing we've discovered is our own star. However, we humans also deserve some of the plaudits in the quest for circular perfection, not simply because we have measured the roundness of the Earth, or other planets, or even stars. The most perfect sphere that we are aware of is also in space, but it is our own creation. We have made spheres of alarming near-perfection, and all in our pursuit of knowledge about the fabric of the universe – more specifically, to test the theories of Albert Einstein.

General Relativity is the model Einstein constructed to describe the fundamental gravitational structure of the universe. According to this theory, spacetime – the actual fabric of space – is dented by large objects with a lot of mass, such as planets. One way – the standard way we get taught at school – to visualize this fairly baffling thought is to think of a single plane through space as a taut rubber sheet. If you place a bowling ball on the sheet, it distorts the fabric and creates a dip. This dent occurs in the third dimension, if the sheet is just two-dimensional. But of course space is already three-dimensional, so the dip of a large body such as a planet dents into the fourth dimension. Massive objects (that is, not necessarily big objects, but objects with a lot of mass: some astrophysical phenomena are immensely dense but not very big, so are massive but small, such as the neutron stars that spherical bastard Fritz Zwicky conceived of) warp the fabric of spacetime.

As a planet rotates in space, General Relativity predicts that it will create a tiny vortex in that dent, like spinning a ball in honey – a phenomenon called frame-dragging. In order to test this, physicists in the 1960s concocted an experiment as simple in concept as that of Eratosthenes 2,000 years earlier, but requiring some slightly more complex kit than a stick and a fit servant.

The idea is straightforward: if you spin a gyroscope, the main axis tries to point in a single direction. On Earth, there are all sorts of forces that slow it down and send it off course, such as gravity, friction and the presence of an atmosphere. But if you spin a gyroscope in space, where there are no external forces, then it should retain its spin perfectly, and never divert from that axis. Gravity Probe B was designed as the centrepiece of an experiment to test for frame-dragging: the idea was to send a gyroscope up into orbit, point it at a star, and see if the effect of the mass of the Earth put a kink in its spin.

The expected tilt, if Einstein was right, would be in the order of 0.00001167 degrees per year – that is, an inconceivably tiny slice of a circle. So the gyroscope had to be unimaginably accurate, and this was achieved by making ping-pong-sized ball bearings that were as spherical as they possibly could be. They were in fact the most perfect objects ever crafted. They were made of fused quartz and silicon, and deviated from a perfect sphere by a *maximum* of . . .

40 atoms. That is considerably less than the depth of the ink on this page. If we were to scale up one of these spheres to the size of the Earth, there would be no hills higher or valleys deeper than *12 feet*.

The perfection in this experiment doesn't end there. These balls (four of them, just in case any single one failed) were suspended in 400 gallons of liquid helium, maintained at a temperature of −271° Celsius, that is, 1.8 degrees above absolute zero. This was crucial for detecting any tilt. The balls were coated in a near-perfect film of the metal niobium, which at that temperature becomes a superconductor. When a superconductor spins, it produces a magnetic field precisely parallel to the axis of spin, and this can be picked up by a wickedly ultra-sensitive device called a Superconducting Quantum Interference Device, or SQUID.

These gyroscope balls were mounted in Gravity Probe B in their liquid helium chambers and sent up to orbit the Earth, pole to pole. Because this orbit also had to be breathtakingly precise for the experiment to work, the launch window was limited to a single second, and after one attempt was scrubbed because of wind, it went up on 20 April 2004, at 4.57 p.m. and 23 seconds.

The spaceship also carried a telescope that was set to point in the direction of a star, IM Pegasi, which is 329 light years away, and visible on a clear night in the Pegasus constellation. The balls were set to spin with their axes pointing directly at the star.

The experiment ran from 28 August 2004 until 4 August 2005. Then began years of crunching the data to see if the axes of the spinning balls had deviated from their original true direction. On 8 December 2010, Gravity Probe B was formally decommissioned, and within a year the team of physicists behind this spectacular experiment published their results. Einstein's General Relativity model predicted that the frame-dragging drift rate for the Earth should be 37.2 milli-arcseconds per year. Gravity Probe B measured the frame-dragging drift rate at 39.2 milli-arcseconds per year.

So while it may be true that nothing is perfect, the nearest thing to a perfect sphere that anyone has found was built by scientists. The science-fiction writer (and also rather good scientist) Arthur C. Clarke once said that 'any sufficiently advanced technology is indistinguishable from magic'. Gravity Probe B was an experiment that was indistinguishable from art – Einstein, perfect spheres, space SQUIDs, Pegasus. As far as we know, it is still orbiting us, around the poles, its mission complete, looping about 400 miles above our fat-around-the-middle planet. Tucked inside its guts are four perfect ping-pong-sized spheres which showed that Einstein was right.

CHAPTER 4

ROCK OF AGES

Now that we've established that the Earth is not quite a sphere, but a bit saggy around the middle, we wish to ask another slightly impertinent question: just how old is this large, craggy rock that we live on?

How did the Earth begin?

Imagine you didn't know about the Big Bang. Imagine you didn't know about a solar system formed from gigantic dust clouds, or about the molten rock slowly sloshing around in the bowels of the planet. If you came to ponder this question without the luxury of modern astrophysics and planetary geochemistry, where would you start? It would be bewildering. This goes some way towards explaining why so many creation stories from the ancient civilizations are completely mad – perhaps none more so than that of the Greeks.

In the creation myth of ancient Greece, the Earth mother, Gaia, was an eternal power drawn out of Chaos, the realm of disorder. She came into being along with Tartarus (the Abyss) and Eros (Love), and soon gave birth to Ouranos, the Sky, whom she also married.

Shortly after that, Ouranos impregnates Gaia and she gives birth to the Titans. These 12 children (six male, six female) spend an immeasurable amount of time doing some outrageously sordid things – things that, frankly, do a good job of helping you forget that their father is also their brother.

Before long, Ouranos tires of their shenanigans and imprisons the 12 in the Underworld. Only his youngest son-slash-brother Cronus is bold enough, at his mother's request, to confront his father, which he does by slicing off his genitals with a scythe and hurling them into the sea (these, by the way, become Aphrodite, the goddess of beauty). There's a bit more incest, and soon enough the Olympian gods arrive on the scene. Things don't get much less confusing with the arrival of Zeus, Hera, Poseidon and all of those capricious deities ruling the Earth, but, by the heavens, they're fun.

The stories of other civilizations and cultures are frequently no more comprehensible. A favourite of ours comes from Finnish folklore. According to the legend, Ilmater, the daughter of the sky, floats for centuries until she finally settles for a 700-year swim in the seas.

A bird notices her and lays seven giant eggs on her knee. She balances these eggs for a few thousand years, during which time they heat up, eventually getting so hot that she cannot

bear it any longer: they fall off and crack in the oceans. The yolks form the Sun, the whites become the stars and the shell becomes the Earth.

There is a pattern to these creation myths. However uniquely bonkers they appear on the surface, they broadly fit into a handful of categories. The chaotic origin is one type (as in the Greek story), where order is drawn out of disorder; the cosmic egg is another, appearing not just in Finnish lore but in Chinese mythology, in Hinduism and in dozens of other cultures. Another type is *ex nihilo* – out of nothing – and this includes not only the Christian myth but also, rather interestingly, the scientific version of creation that we now accept as being the most verifiable, the Big Bang.

How long ago did the Earth come into existence?

In all creation stories, Time is a character that doesn't seem well connected to our own perception of the passage of time.

Few people talk about the Big Bang happening on a Tuesday afternoon. Days, weeks, months and hours have no meaning at the origin of the universe. It's as though we're almost all comfortable with agreeing that these events happened 'long ago' and leaving it at that.

There are some, though, who don't treat their creation stories as metaphors and myths, but instead look within them for evidence of the true age of the Earth. We're talking specifically here about the biblical creationists: the loud but fringe cousins of very much more sensible Christians. They don't interpret the poetic, sometimes nutty, often genocidal but at other times beautiful stories of the Bible. Instead, they assert that it is literally true, word for word: Jacob wrestling the angel, Job in the belly of a whale, Noah and the flood, the life and times of Jesus in the New Testament.

As an aside, in case you're wondering, the dimensions of Noah's Ark are provided in the Bible, making this gopherwood ship a fair bit smaller than the *Titanic*. Attempts have been made to work out if all the animals could fit on such a vessel, and while there is room for a bit of interpretation in the calculations, from a zoological perspective three rather large problems immediately spring to mind: (1) sticking a bunch of apex predators in a small space with lots of tasty prey is not a great idea if you want them all to go forth and multiply; (2) repopulating a planet's worth of creatures from a stock of just two individuals would be a genetic disaster for any species – the profound levels of resulting inbreeding would render it functionally extinct; and (3) can you imagine the amount of dung they would collectively produce? It would be a heap of biblical proportions.

We digress. Creationists believe that the Bible is a literal record. Even the largely overlooked story with a talking donkey,[*] and that time God sent a couple of bears to kill 42 children after they laughed at a bald man.[†] And even, crucially, the creation of the universe and everything in it in six days, the seventh being for God to have a well-earned day off, maybe watch some Netflix.

Part of the creationist philosophy is that 'long ago' wasn't nearly as long ago as other religions, or science, might suggest. Like scientists, biblical creationists hanker after precision, and many of them stick steadfastly to a date that is so precise that it warrants proper scrutiny. According to one eminent seventeenth-century cleric, on the basis of a literal reading of the Bible, the events of Genesis Chapter 1, where God decreed 'Let there be light', happened in 4004 BCE.[‡] This means that as we write these words, in the year 2021, the Earth, and indeed the universe, are 6,025 years old.

This, we hope it won't surprise you to learn, is not correct. The evidence for it being not correct is so bountiful it's almost not worth going into. You have held stones in your

---

[*] Numbers 22:21–39. Balaam is out riding his donkey when the donkey notices an angel standing in the road and swerves to avoid him. Balaam whacks the donkey. This happens twice more, and after the third beating the donkey pipes up and says: 'Dude, why are you hitting me?'

[†] 2 Kings 2:23–4: 'Then he went up from there to Bethel; and as he was going up by the way, young lads came out from the city and mocked him and said to him, "Go up, you baldhead; go up, you baldhead!" When he looked behind him and saw them, he cursed them in the name of the Lord. Then two female bears came out of the woods and tore up 42 lads of their number.'

[‡] We used to say BC, as in 'Before Christ', and AD, as in 'Anno Domini', the Year of our Lord. Nowadays, science has adopted the abbreviations CE, meaning the Common (or Current) Era, and BCE for Before That. For convenience's sake, it is agreed that the Current (or Common) Era begins in the year 0 on the BC/AD scale.

hands that were forged in the heat of the Earth millions of years ago, even billions. The dinosaurs lived for 170 million years, and all (well, all the big ones) perished 66 million years ago. Animals that live in very cold waters grow very slowly, and some estimates suggest that there are Atlantic glass sponges alive today that are up to 15,000 years old. To put that another way, these animals were alive before humans had even domesticated pigs.

Winding the clock back 6,000 years, you'd find yourself sitting in an era we refer to as the Neolithic. Agriculture was blossoming globally; ploughs were being developed in Europe; young children were being fed using clay baby bottles; maize was being cultivated in the Americas, copper forged in Egypt; and dairy farming and cheesemaking were part of life in Africa, in the Middle East and in Europe. There was plenty of human activity already going on, basically. Six thousand years ago was also the beginning of history, the point at which we first started recording things in writing.

Biblical creationists are wrong. Very, categorically, comically wrong. And there are myriad reasons why we know that the Earth is emphatically, colossally, magnificently, ever so muchly much much more than 6,000 years old.

It might surprise you, therefore (it certainly surprised us) that right here in this science book – written by two people who have spent their lives espousing the importance of science – we wish to defend this number. There are few obvious clues to how old things really are. Trees' annual growth can be measured in the rings of new tissue they lay down, so we can use the rings to establish their age – it's called dendrochronology – but the oldest tree is a mere 4,850 years old, give or take; it's called Methuselah after Noah's grandpa, who the Bible says died at the age of 969, which again suggests that Time is a flexible character in such tales. So, in the Middle Ages, scholars relied on the most trusted sources of information they had. In Europe, a predominantly Christian continent, the Bible was considered the infallible source of such facts, and the date of creation was an important and scholarly topic. Johannes Kepler, the Venerable Bede, even Isaac Newton all made serious attempts to pinpoint the moment of creation, and all hovered around a time of about 6,000 years ago. In particular, one date stands out, and has become a favourite among present-day creationists. It is the date for the creation of the Earth revealed by an Irish archbishop called James Ussher: 23 October 4004 BCE.

The six-millennium Earth

Ussher was born into a wealthy Dublin family in 1581, and went to university at the age of 13 (yes, he was certainly smart, but it wasn't that unusual at the time). He shot up the ecclesiastical ladder, and was appointed to the *Harry Potter*-esque-sounding post of Professor of Theological Controversies at Trinity College Dublin at the age of 26. This ascent continued all the way to the top job in the Irish Church in 1625, when he was appointed Primate of All Ireland. It will never not amuse us that one of the most senior rankings in the Christian hierarchy is also the taxonomic name for monkeys and apes, but that is a trivial aside.

These were troubled times in the British Church, with the country heading towards the political and religious turmoil of the Civil Wars. In 1649, Charles I was beheaded in London at the Palace of Whitehall; Ussher watched from a nearby roof, but passed out as the axe fell.

From this point on, Ussher busied his considerable mind and dizzying intellect with the task of working out when the universe began. A year after Charles's execution, he published his masterwork, *Annales veteris testamenti, a prima mundi origine deducti – Annals of the Old Testament –* in which he methodically set out his calculation of the origins of the world (and indeed, everything else).

Ussher made some assumptions, which were typical for this type of project at this time: for instance, that creation occurred on a Sunday. This, of course, was because God rested on the seventh day. The day of rest for Jewish people is the Sabbath, which is on Saturdays, so creation must've occurred on the previous Sunday for God to be able to kick back on the Sabbath.

He also assumed that the Earth came into being at precisely 6 p.m. the night before that crucial Sunday, because that is when the day begins according to the traditional Jewish calendar. It was also the traditional Jewish calendar that gave rise to Ussher's belief that creation happened in the autumn, as the Jewish new year occurs at the autumn equinox, when day and night are exactly the same length.

The eagle-eyed among you may have noticed that we now mark the equinox on 21 September, not 23 October. That's because the Gregorian calendar we've used since the sixteenth century is up to ten days behind the Julian calendar introduced by Julius Caesar in 45 BCE and used by Ussher. The reason for this mammoth discrepancy boils down to what a year actually is. We're going to need you to strap in for this, because it's a wild ride.

The old-school Julian calendar worked on the basis that one terrestrial orbit around the Sun takes precisely 365.25 days. We can't start days at 6 a.m., a quarter of the way through 1 January, so we round the quarter down three years in a row to 365 days, and round up to 366 every fourth year – a leap year.

So far, so familiar. But the length of one Earth orbit of the Sun is actually 365.2425 days. That difference makes the Julian year about 11 minutes too long. Over the centuries, this discrepancy had built up, and the Julian calendar was days out of line with the Earth's orbit. We deal with this today by dropping the leap year in the first year of every century that is not divisible by 400, meaning that 1700, 1800 and 1900 were *not* leap years, nor will 2100, 2200 and 2300 be. But when Pope Gregory XIII spotted the anomaly and fixed it in 1582, he made the initial shift to the calendar that bears his name with a rather more radical step: he erased ten days from history, so that the day after Thursday, 4 October was Friday, 15 October. It is not clear what happened to those poor souls who had a birthday in those vanished days.

Still with us? Because that doesn't yet put Ussher's date of 23 October at the autumn equinox, and therefore the moment of creation. But Ussher used what's known as the *proleptic* Julian calendar. Before Caesar regularized leap years, they still needed to happen to prevent the annual calendar getting way out of kilter with the seasons; but the additional years were enforced erratically, sometimes occurring every three years, other times just when required. It was only in 8 CE that leap years were officially introduced every four years, that is, made quadrennial. So anyone working out dates before that had to take into account the variability of leap years. We told you Ussher was methodical.

THE TIMES THEY ARE A-CHANGIN': A QUICK GLOSSARY OF WESTERN CALENDARS

The Julian calendar: Designed under the auspices of Julius Caesar in 45 BCE and regularized in 8 CE to recognize that the orbit of the Earth around the Sun is 365.25 days, so that every fourth year we round up with a leap year of 366 days.

The proleptic Julian calendar: When historians used the Julian calendar to work out dates before 8 CE, they had to include the fact that leap years hadn't been regularly enforced until then.

The Gregorian calendar: Pope Greg XIII and his team worked out that Caesar had fractionally overestimated the length of one Earth orbit, which meant that the Julian calendar was creeping ahead of the annual orbit. So they removed ten days in October 1582 – just erased them from history – and also removed three leap years every four centuries.

The Jewish calendar: Same as the Gregorian, but the new year starts at the autumn equinox.

Clear?

So, by fiddling with the various calendars and their weird temporal anomalies and fixes, Ussher had nailed down the date of creation to the evening before 23 October. Now all that remained was working out *which* 23 October. The general presumption at the time was that the Earth was approximately

6,000 years old, on the basis of a passage in 2 Peter 3:8: 'But, beloved, be not ignorant of this one thing, that one day is with the Lord as a thousand years, and a thousand years as one day.' Ussher, like most other scholars of his time, went for a straightforward interpretation on this one: it took God six days to make the whole thing, and a divine day is metaphorically equivalent to a thousand years, so the Earth must be 6,000 years old – 4,000 years from Day One to the birth of Jesus, and 2,000 more after that.

Ussher calculated that the Earth was 6,004 years old; the extra four years were drawn from the historian Josephus, who had attempted to correct another earlier chronological error. The switch from BCE to AD was marked by the birth of Jesus, but Herod's death coincided with a lunar eclipse, which Josephus calculated had occurred in 4 BCE. For the Gospel of Matthew to be correct – remember that the Bible is considered inerrant in this period, but doesn't have any dates in it – Jesus must have been born before Herod's death, because he escaped the king's decree that all male newborns should be killed. Hence, Ussher figured, Jesus was actually born in 4 BCE: and that, in turn, means that at 6 p.m. on the evening before 23 October, 4004 BCE, God glanced at his digital watch and – boom – made the universe.

You may now take a day off and watch some Netflix. Ussher's calculation, along with those of all the other chronologists of the time, was in keeping with the scholarly belief that God was an astronomer, and that a logical and rational approach to dissecting the evidence to hand would reveal His perfect universal mathematical architecture. The main problem with this endeavour was not Ussher's methodology; it was the evidence. The title of his book is *Annals of the Old Testament*, but it turns out that the Old Testament is not a science book. It's not really a historical text either. It's easy to mock Ussher's chronology, for it is littered with assumptions and relies on the Bible being factually correct, which most of us no longer believe. But by the standards of Ussher's day, biblical accuracy was a given, and his scholarship was precise, logical and thorough. Yes, he may have been more than five orders of magnitude wrong; but he was working with the accepted wisdom of his time, and applied it rigorously to come up with this number. What *is* madness is to cling to this methodology and this date in the twenty-first century.

After Ussher's time, science eventually became formalized into a set of methods and tools for testing observed reality, and stopped relying on doctrine or edicts handed down by authorities. Our best estimate today is that the Earth is actually around 4.5 billion years old – we will come to the evidence for that in a minute – meaning that Ussher was wrong by a factor of around 760,000. As error margins go, that is a whopper. It's equivalent to claiming that the distance from London to Birmingham is not 120 miles, but in fact 10 inches. But figuring out the age of ancient things takes a lot more tricks and tools than measuring the distance between two places on Earth. Honestly, can you demonstrate that a rock is millions of years old, and not merely a few thousand?

Creationism faded into relative obscurity in the years when science blossomed. By the time we get to the first half of the nineteenth century, Charles Darwin was working out his theory of evolution by natural selection, and he figured that in order for it to be correct, the Earth had to be millions of years old. On his travels aboard the *Beagle*, he encountered a menagerie of beasts, some of which seemed to have characteristics similar to others from which they were separated by unswimmable oceans. He needed continents to move over time – and indeed they do, though it would be another century before plate tectonics was described accurately.

But it was in the soils of the South Downs that Darwin found an Earth older, much older, than the estimates of previous centuries. Darwin knew that the fossil record was extremely incomplete, because the process of fossilization requires particular conditions which are not very likely to coincide: you need the right ground, the animal to be covered quickly so that it doesn't get eaten, and all that land to stay pretty stable until we dig it up a few million years later. Even though we now have millions of fossils, in Darwin's day they did not show the perfect continuity of gradual change required to demonstrate one species slowly transforming into another, which was a cornerstone of his theory. So he used the work of some of his geologist colleagues, particularly Charles Lyell, to show that the missing transitional animals would have lived over periods of hundreds of thousands, if not millions, of years. The Weald in southern England is a huge geological basin, with chalky ground just below the topsoil, stretching from just south of London all the way to the Channel coast. The model for how it formed was that it was once a giant mound called an anticline, a mountain pushed up from the churning Earth beneath; over huge swathes of geological time the top was eroded, flattening its peak and exposing the very ancient chalky bottom below. Darwin looked at erosion rates for different types of rock and concluded that the Weald Anticline had taken a whole 300 million years to erode to its present status as a basin, including the relatively flat lands that lay beyond his back garden at Down House in Kent. This gave him a much longer period in which animals could transform into other animals, according to his brilliant and completely correct theory of evolution.

But though he was right in principle, he was also well out for the age of the Earth – out by a multiple of 15 – which, while considerably closer than Ussher's chronology, is still not an error margin you'd want if you were setting your watch.

The most accurate and recent date for the age of this old rock we live on comes from the fusion of geology with chemistry and nuclear physics, and that all starts with Ernest Rutherford.* Along with his work showing that the atom was made of electrons, neutrons and protons, Rutherford also spotted the fact that some atoms have more neutrons in their nucleus than others, which makes them heavier versions – isotopes – of the same element, and that these are generally less stable. This instability is what makes them radioactive: as they break down, they release particles (or gamma rays). Rutherford also noticed that elements decayed at a predictable rate, and that this rate was unique to each radioactive element. After an experiment establishing that it took 11½ minutes for a particularly rare sample of thorium (named, by the way, after the goat-eating Norse god Thor) to decay by 50 per cent, he came up with the concept of 'half-life'.† Because lots of elements have radioactive versions, and these are mixed in with the stable type, Rutherford's discovery offered the prospect of a kind of nuclear clock. If a rock contains atomic matter in elements that have known radioactive isotopes, we can assume

---

* No relation. To either of us.
† There are 30 different isotopes of thorium, all with 90 protons but with between 118 and 148 neutrons. All are unstable, but the lengths of half-lives for the different types ranges from under ten minutes to 14 billion years – which is approximately as old as the universe.

that these atoms were incorporated in the rock when it was formed, and not afterwards. If we know what the typical ratio of radioactive atoms to normal ones is in that element, then we can measure the number of radioactive atoms in a sample of the rock today, and, knowing the half-life, calculate when the rock was formed. This is also the process by which we date specimens of things that were once living. A proportion of carbon is radioactive; we incorporate this carbon into our bodily tissues by eating, and as long as we are alive, that proportion stays steady. When we die, we stop eating, and stop incorporating carbon, so the clock starts ticking as the radioactive carbon decays, and the proportion goes down. This is called radiocarbon dating – but because the half-life of carbon is a measly 7,530 years, this technique only works for specimens that kicked the bucket up to about 50,000 years ago. If they're older than that, there isn't enough radioactive carbon left to arrive at an accurate date.

One version of this widely used technique tells us the age of the Earth with what in other hands might become a bit of cheap jewellery: zircon, an affordable alternative to diamonds. Zirconium silicate crystals are almost perfect cubes, and they have a uniquely useful feature: when they form, they can incorporate an atom of uranium into the middle of the crystal structure, but they cannot trap lead atoms.

A small proportion of uranium is radioactive;[*] this is U238, which decays and becomes a form of lead, Pb207, with a half-life of 4.47 billion years. Because zircon can only trap

---

[*] Uranium, by heavenly design or scientific coincidence, is named after the planet Uranus as they were discovered in the same decade, and Uranus is the Roman version of Ouranos, the castrated Greek sky-god.

uranium when the crystal is forming, once a radioactive atom of U238 is trapped, the clock starts ticking. Zircon crystals are mighty tough, capable of surviving crushing forces of geological metamorphoses over billions of years.

So when we find zircon crystals in rock formations, we can stop the clock. We know that any lead (Pb207) found in zircon began its entombment as U238, and by working out the proportions of Pb207 to uranium in a zircon crystal deposit, we can tell exactly how old it is. The oldest deposits found so far are from the Jack Hills in Western Australia, and they give us a time of 4.404 billion years since their formation.

We get an even older date applying a similar method to meteorites assumed to be extra material from the Earth's formation, when bits of floating debris left over from the birth of the Sun crunched together in orbit, and stuck – a process called accretion. Analyses of these meteorites have homed in on radioactive lead, osmium, strontium and a whole load of other heavy metals, and they all come in at around 4.5 billion years, with error margins of a few tens of millions of years.

So there it is. Our planet is about 4.5 billion years old, which is about a third of the age of the universe itself. Finding the answer to that question has occupied the minds of some of the greatest scientists of their ages, and though their answers were all different, and differently wrong, all of them have ended up in the ground together. Along with Isaac Newton, Charles Lyell, Charles Darwin and Ernest Rutherford, James Ussher was buried in Westminster Abbey. His tomb bears the legend 'Historian, Literary Critic, Theologian. Among saints – most scholarly. Among scholars – most saintly', which, let's face it, is a pretty good homage to a very clever primate.

CHAPTER 5

A BRIEF HISTORY OF TIME

This is a story that begins with a cable. It's a very particular cable, hidden under the streets of London, buried deep in the earth. It would seem innocuous enough if you stumbled across it – just a fistful of fibres, coated in black, running from a little beyond the grounds of Henry VIII's Hampton Court Palace south-west of London to Docklands on the shores of the Thames further east. Hardly anyone knows that it's there. Even fewer people know *why* it's there, or how a trillion dollars that disappeared into the ether in the space of just over 30 minutes meant that the world – as we knew it – could no longer carry on without it. It's a cable with unimaginable sums riding on the numbers that pass through it, but its job is beautifully simple. It serves to answer only one question, with more accuracy than almost anywhere else in the known universe.

What time is it?

Of all the clocks at your disposal, you probably have an idea of the one you trust the most. Maybe it's your mobile phone, set by a remote signal; or your precision-engineered Swiss wristwatch. It's probably not the clock on your oven, which is still showing the same time as when it was installed, and can only be reset with instructions you've almost certainly lost.

In just the same way, humanity has a hierarchy of clocks, and some are more trustworthy than others. None of them are on ovens. Pendulum clocks are more dependable than sundials; quartz clocks are more reliable than pendulums. That makes it fairly straightforward to test the accuracy of a clock. You simply compare it with one higher up the hierarchy – one that you trust to be more accurate – and it'll soon be obvious if the seconds aren't keeping pace with each other.

If there is a hidden hierarchy of clocks, you might assume there must be one that sits at the top of the pile – a timepiece that sets the standard for all other clocks to follow. But time is not like mass, with its standard units (until 2018, the definitive kilogram was a platinum cylinder kept in a vault in Paris); a single unit of time – such as the second – can't be defined in the same way. You can't build an archetypal second, place it in a glass jar and lock it in a cupboard, or hold a second in your hand. So, to answer the seemingly simple question of what time it is, we must first know who, or more pertinently *what*, is the ultimate arbiter of time, and address the small matter of what time actually means.

What is a second?

This should be easy enough to answer. After all, the Earth rotates once about its axis every 24 hours, each hour of which is 60 minutes long, each minute containing exactly 60 seconds. If you want to measure how long a second is, it should simply be a matter of pointing a telescope straight up at a star in the sky and waiting until the same star comes back around to the same spot the next night – that is, an exact day later. If you divide the time that's elapsed by 86,400 (the number of seconds in the day), then you should end up with precisely the length of one second.

It seems like a good idea that the Earth itself, rather than a physical clock, should be the ultimate timekeeper. But even in antiquity, astronomers had an inkling that the Earth's spin didn't make for a reliable stopwatch.

Sundials can give you a pretty good estimate of how far through a turn the Earth has spun on its axis. But as soon as anyone tried to set up another timepiece to run alongside a sundial, time started to warp. They tried giant hourglasses and water clocks (using drops of water instead of sand); they even tried candle clocks (where the burning wax and the shrinking candlestick indicate the passage of time), but always came to the same conclusion: human-made devices and sundials couldn't keep time with one another. Sometimes they would run ahead of the sundial, sometimes they would fall behind.

Ancient Babylonian scientists had known about the irregularity; so too had Ptolemy, the second-century Roman mathematician and astronomer, who had even gone so far as to calculate the correction needed between the human-made clocks and the celestial. But it wasn't until the Dutch physicist, mathematician and amateur horologist Christiaan

Huygens invented the pendulum clock in 1656 that the world had a device that could be used to investigate the relationship between sundial-time and clock-time with precision. Huygens's pendulum clock could keep time to within a few seconds a day, which made the whole job much easier (and much less messy) than trying to make precise calculations sitting next to a sundial with a ruler and a half-melted candle. Using his newly invented clock, Huygens created a series of tables comparing the discrepancies between the two timepieces across the course of an entire year, offering proof that something was clearly awry: because, as he found, pendulum clocks and sundials disagree on what time it is, but the disparity between them isn't random.

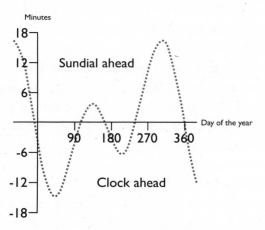

The amount by which clocks and sundials disagree ebbs and flows across the course of a year, and Huygens noticed that each and every year seemed to show the exact same pattern. The human-made clocks were ticking away in time with each other; it was the sundial that appeared to be cranky. True, the clocks of the day were not exactly Swiss pocket watches, but the Earth was worse.

The reason why lies in the celestial realm. There are tiny but measurable fluctuations in the Earth's rotation during the course of a revolution around the Sun. As our planet orbits its star, the gravities of other planets make slight dents in our orbital path, which means that sometimes we spin slightly faster, sometimes slightly slower, like a waltzer at a fairground. The impact of these gravitational lumps and bumps points to a pretty heavy timekeeping concept: a day is not actually 24 hours long.

A day in September – if you define it by the time that elapses between two consecutive points when the Sun is highest in the sky (noon to noon) – is almost a full half an hour shorter than one in February.

Look at the face of the pendulum clock instead of the sky, and then midday to midday is indeed 24 hours irrespective of the month; but the Earth may have under- or over-rotated in that 24-hour period, depending on the time of year. You won't notice it happening, because day by day it's only out by a handful of seconds, but it's enough to mean that any human-made clock can't possibly keep pace with the sundials, because celestial timekeeping is irregular.

In earlier ages no one needed to know when it was precisely 11.38 a.m. Our daily routine was much more closely wedded to the rhythm of the Earth. But as we began to traverse the globe, it became more and more essential to know what time it was (see box on page 118). If clocks were going to be useful (which they had to be, because not everyone had a cockerel to tell them to wake and an astronomer to tell them when to serve afternoon tea), this problem needed a robust workaround.

So, in the seventeenth and eighteenth centuries scientists had the idea to redefine a 'day' as the *average* time it takes for the Earth to spin on its axis (which is 24 hours), measured across an entire year, rather than the time between consecutive high noons (which is 24 hours-ish, plus or minus a quarter of an hour). Since the new day was based on an average, they'd call it 'Mean Time'; and since the Brits claimed to rule the waves at this point in history, they got to name it after their most notable observatory: Greenwich Mean Time.

The wobbly world

Once Greenwich Mean Time had been established, everyone could relax. No more puzzled scientists accidentally burning themselves while taking observations from candle clocks, no more ships' captains wondering where they were going. Afternoon tea was served at 4 p.m. promptly the world over.

The pendulum clock reigned as the supreme timekeeper for more than 275 years. Those stationed in London, set by astronomical observations and running to Greenwich Mean Time, remained the temporal authority. But then, as the twentieth century rolled on, the arrival of a couple of newly invented timepieces reignited the debate all over again.

Did the world finally have an answer to what time it is? Not even close.

First came quartz. In the late nineteenth century, Paul-Jacques and Pierre Curie (Marie's brother-in-law and husband respectively) discovered that if you crush a small piece of quartz crystal it will produce a tiny electric charge. It's an effect that works in reverse too: apply a small electric field across a piece of quartz and it will react by contracting and releasing a small electrical charge. It makes quartz fantastically useful – during the First World War, for instance, the French made a submarine detector based on this very property* – but it wasn't until the 1920s that quartz's potential for clocks was realized.

The contraction and release of quartz when electricity passes through it happen at a very predictable frequency. These oscillations could effectively replace the swinging back and forth of a pendulum, meaning that the number of vibrations per second could be used to keep track of time much more reliably than a mechanical clock could ever manage.

But as soon as scientists compared their shiny new quartz timepieces to the Earth's rotation, they realized – to their dismay – that there was a whole other layer of time-based noodlery to worry about. They spotted a discrepancy between the two, and became suspicious of their clocks all over again. The predictable speeding up and slowing down of the Earth they already knew about; but even after taking that into account, with these ultra-reliable quartz clocks that were using molecular structure as a pendulum, the Earth still wasn't keeping pace.

Quartz clocks and the Earth disagree on what time it is. But the amount by which they disagree is not neat and predictable. There's another layer of irregularity – something that can't be fixed by using Greenwich Mean Time – that is quietly warping and distorting the length of a day on a daily basis. We just didn't realize this layer was there until we had invented timepieces accurate enough to reveal it.

---

* This was an early form of sonar. The idea was to send out a high-frequency pulse into the ocean; when it bounced back, the quartz could detect the vibrations in the water. The echo would squish the quartz, and it would respond by giving off a tiny electrical signal. The more quickly the echo came back, the closer the object in the water, so sending off lots of pings could tell you when something ominous was approaching.

YOU DON'T KNOW WHERE YOU ARE UNLESS YOU KNOW WHEN IT IS

In 1707, the War of the Spanish Succession was raging, with the powerhouses of Europe scrabbling around to claim the territories released from the Spanish Empire after the heirless death of Carlos El Hechizado – Charles the Bewitched. A British fleet set sail from Gibraltar, bruised from battle with their sworn enemies, the French, and headed home for Portsmouth. The weather was bad, and on the last leg of their journey it turned treacherous and forced their ships way off course. The various captains recalculated their position and believed themselves to be sailing safely off the coast of Brittany in France, when in fact they were fatally headed directly towards the shallow waters and rocky outcrops of the Isles of Scilly. Before anyone realized their mistake, four ships ran aground and as many as 2,000 souls were lost. It was one of the worst naval disasters in British history.

The mariners' mistake had been, at least in part, in calculating their longitude. It's relatively easy to work out how far north or south you are – latitude – because all you need is the calendar date and the angle of the Sun above the horizon: the Sun gets higher in the sky during summer in a predictable and consistent way that means you can just look up your position. But it is almost impossible to make a similar simple astronomical calculation for how far east or west you are when at sea. To do that, you need to have some way to tell the time, accurately.

It's fairly simple to work out the local time. The Sun reaches its maximum point – the zenith – at noon, so you can watch the sky over the course of a day to pinpoint that moment. But it's only if you also have a clock that tells you the Mean Time in Greenwich at precisely noon, locally, that you can work out the time difference from London; and it's that difference you need to know to work out how much of the Earth's rotation stands between your position and home – that is, how far east or west you are of the Greenwich Meridian: your longitude.

Clocks on ships at this time just couldn't do the job. The pendulum clocks of the day would keep time well on land, but weren't capable of functioning effectively on a swaying ship out at sea. Changing temperatures, too, meant the metal inside the clock would bend, expand and contract, mangling the speed at which the cogs turned. Briny humidity wreaked havoc with the gears. All of these mechanical shortcomings made for disastrous miscalculations in longitude.

Such was the desire to rule the waves, and so serious was the risk to life if you couldn't tell the time at sea, that Spain, the Netherlands and France all offered big cash rewards in the eighteenth century to anyone who could build an accurate, seaworthy timepiece. But it was in Britain, where in 1714 the government promised a staggering prize of £20,000 – around £4 million in today's money – that success was finally realized. Although the prize was never awarded in full (politicians are surprisingly good at finding excuses not to give out free money or honour their pledges), John Harrison, a British carpenter and clockmaker, is recognized as the man who engineered the solution. By 1761 he had managed to build an ingenious oversized pocket watch called H4, which included a bimetallic strip that regulated the mechanism as it changed size with temperature. It lost only 5.1 seconds over a voyage of 81 days in its first trial at sea.

That, of course, is positively pitiful compared to a modern-day quartz watch (which can track mean solar time within ten seconds a year); but the legacy of Harrison's longitude solution lives on. By the nineteenth century, it was considered unthinkable to travel at sea without an accurate chronometer, and even now, however you navigate on Earth – using GPS or good old-fashioned sextants – you are fundamentally reliant on knowing the time.

The need to tell the time was an essential step in maritime history for one simple reason: you don't know where you are unless you know when it is.

All this means one thing: not only is any particular day not 24 hours long; an *average* day is not 24 hours long either.* As a timekeeper, the Earth is absolutely all over the place. The precise time it takes the Earth to make one whole rotation is *totally* unpredictable.

There is a reason for all this: once you're measuring at the level of microseconds, winds and atmospheric pressure can be enough to subtly change the precise length of a day on our planet. We now also know that the Moon is slowly spiralling

---

* By a 'day', here, we mean the time for one Earth rotation. Which was what it meant in the days before pendulums. Then 'day' changed to mean 24 hours, which is not the same as the old day, but is the day that we use today. It's probably best not to think about this too hard.

away from us, slowing down the Earth on its axis (see box over the page). Even the heavy liquid outer iron core sloshing around almost 3,000 kilometres below our feet can have an impact on how quickly we're spinning. Wind, liquid cores, gravity – they're all chaotic processes that you can't correct for with mathematical tricks.

Trying to set your clock by the Earth, as we have done for most of history, is a waste of time.

There was only one thing for it. Ditch the Earth altogether and stick to reliable, hard-working human-made machines to tell the time. This was particularly appealing after the invention of the atomic clock in the 1940s. After all, the fundamental properties of matter are just a tad more dependable than the speed of a spinning rock wobbling through space.

The pendulum within an atomic clock is an atom of caesium, which, when it resonates, does so at the same frequency as every other atom of caesium in the universe. At the atomic timescale, every second is identical to every other.

The most accurate atomic clock ever built loses less than a second every *15 billion years*. This, by the way, is more than a billion years longer than the age of the universe to date, so ought to satisfy any demands for timekeeping precision.

In 1972, the International Bureau of Weights and Measures, just outside Paris (the same place that was home to the archetypal kilogram), switched to atomic time as the new global standard. It remains the official worldwide clock-setter, regularly taking the read-outs of around 70 different atomic clocks stationed around the world and bringing them together to calculate the official global – and by extension cosmological – time (known as Universal Coordinated Time, which is confusingly abbrievated to UTC). Note what has

CORAL TIME

We know the Earth is slowing down in part thanks to coral.

Corals are colonies of marine animals that secrete calcium carbonate to form their protective homes, which build up to form reefs. The individual polyps grow their exoskeletons in microscopic layers, which means that if you cut them open, you can see the difference in their growth pattern between the changing seasons – just like counting the rings on a tree trunk. Look very closely, and you can even detect the tiny hallmarks of day and night.

In 2018, two American scientists, Stephen Meyers and Alberto Malinverno, realized that if they counted up all the layers across the seasons for an entire year, they could see in the coral's growth patterns precisely how many days and nights there are. They tried this on the rings of a fossilized coral from 430 million years ago and noticed something very peculiar about the calendar. The coral showed that the number of days in a year was not 365, but more like 420. That's a lot more sleeps between birthdays for your average trilobite.

Meanwhile, fossils from the Devonian period (between 419 and 359 million years ago) reveal that, back then, there were 410 days in a year. This is long before the dinosaurs appeared (they turned up over 230 million years ago), and it goes to show that the number of days in a year has been constantly changing over time. Meyers and Malinverno calculated that the length of a day has been increasing by one-74,000th of a second per year for the last several hundred million years, and this is set to continue for the foreseeable future – 'foreseeable' here meaning a few billion years. So worry not, time is on our side.

just happened there: the official universal time as decreed by physicists operating the most accurate timepieces in history is a *calculation*, not an absolute tick followed by an inalterable tock. You can be forgiven for not correcting your oven clock. Every single clock in the world is wrong.

It's a decision, not a clock, that sits at the top of the time tree. Once a month, the team in Paris send out an email issuing the information required for each nation's Master Timekeepers (yes, that is their real job title) to synchronize their own clocks to the agreed consensus time: the National Physical Laboratory in the United Kingdom, the Beijing Institute of Radio, Meteorology and Measurement, the United States Naval Observatory. In turn, each Master Timekeeper communicates the time to those below them in the hierarchy, sort of like an atomically precise town crier, shouting at satellites, the BBC pips and your mobile phone.

But there's just one more tiny thing, an issue that comes with aiming for perfection – a sneaky little problem that crops up when we set our clocks by a timepiece that keeps perfect seconds.

If we ditch the Earth entirely, and don't bother to set our clocks by syncing them with its rotation, then the days will start slipping. By the time we reach the end of the twenty-first century, noon will be out by almost a minute. Eventually, far enough into the future, an Earth day will no longer line up with a calendar day. Clocks will provide nothing but confusion, high tea will be promptly served in the middle of the night, and owls will deafen us with incessant hooting. The connection between time and what we consider time to mean will vanish.

The workaround comes in the form of leap seconds – some genuine free time, given to us by the official time lords, that you probably won't have noticed. The leap year was adopted to account for the fact that the Earth's orbit around the Sun is 365.2425 days (as explored in Chapter 4), and in the UK there is a somewhat archaic tradition that 'permits' a woman to ask her beau to marry her on 29 February rather than waiting to be asked. The leap second offers little opportunity to do the same, though you could probably blurt out the words 'MARRY ME' if you were really that bothered. But you'd have to be really on your game here, as extra seconds are added to our clocks at irregular intervals to give the Earth a chance to catch up and to ensure that the gap between planetary time and calendar time never grows too large.

Not everyone is a fan of this system. These occasional jumps make time discontinuous, and that can have complicated and serious implications. It means you can't look forward to the future, or back to the past, and say precisely what time it will be, or was, without referring to a table of leap seconds. None of this matters for most of us, but it potentially poses a real danger for any system that relies heavily on a perfectly precise measure of time. Air traffic control, for example, could all go horribly wrong whenever a leap second occurs, with costs that hardly bear thinking about.

But among the strongest opponents to the leap second are bankers, for whom time is literally money. Which brings us back to that strange black cable running quietly under the streets of London.

Gone are the days of red-faced shouty men waving their arms around in a Wall Street pit, buying and selling shares. Those guys are way too slow. The stock market is now overseen by people, but run by computers, algorithms that can execute ten complete deals in the time it takes an old-fashioned flesh-and-bone trader to blink one of his boringly slow biological eyes.

When it comes to making money, speed matters. The fastest of all, the Usain Bolts of the City, are high-frequency traders, some of whom employ a very clever (sneaky) approach (trick) to interact with (prey upon) other, slower people in the market.

It works something like this.

Imagine you visit a convenience store, stand at the counter and ask the shop assistant for one of the very nice-looking bottles of whisky on the shelves behind him. Unbeknown to you, a high-frequency trading algorithm is listening in and has overheard your order. In the time it takes you to look down at your purse to get out your credit card, the algorithm has zipped past you and bought up all the remaining bottles of that very same whisky on the shelves. Suddenly that makes your purchase – the order you've already placed – seem like a rare commodity. You look up from your wallet to see the price has changed, shot up from what it was a few moments ago. You can't back out now – you've already placed the order – so you pay up, take your bottle and walk away clutching your purchase, having spent more than you intended and feeling a little stung by what just happened. The algorithm can then promptly sell its whisky at an inflated price, piggy-backing on the demand that you just created. The algorithm didn't make the whisky, want the whisky or drink the whisky. It just briefly owned it across the period when it became valuable, as a result of you simply wanting a decent drink.

Of course, we're not talking about whisky here, or convenience stores. These algorithms are now a dominant force in the financial markets, buying and selling stocks and shares

THE WOMAN WHO SOLD TIME

This is Ruth Belville, also known as the Greenwich Time Lady.

In the 1800s, her father inherited a beautifully crafted pocket chronometer, originally designed and made for the Duke of Sussex. In 1836 the Belvilles started a business based on having a portable clock that was more accurate than most people's own timepieces. The service they provided was to adjust people's clocks to the correct time. Right up until the outbreak of the Second World War, Ruth would travel to Greenwich Observatory and set her watch to Greenwich Mean Time, before spending the day visiting her clients and selling them the time by letting them look at her watch. She came under threat when a competitor, St John Wynne, started a business selling a telegraphic time signal – effectively cutting out the middle-

woman and providing more accurate time straight to the customer. Wynne ran a smear campaign to discredit her, and accused her in *The Times* of 'using her femininity to gain business'. She held firm, and her operation continued to thrive, despite reporters sniffing around for juicy gossip. In the end, she graciously said that Wynne's tactics had merely given her free advertising.

and futures. And they're doing it all blisteringly fast – making thousands of trades in tiny fractions of a second.

In 2013, the news media company Thomson Reuters accidentally published a story 15 milliseconds before it was supposed to. The article contained some US manufacturing data that was of great interest to traders, which is why – to make it fair to everyone – it was supposed to go out at exactly 10:00:00 a.m. on the dot.

But Thomson Reuters' clocks were wrong. They weren't out by a margin that a clock-watching human could notice, but it was enough to make all the difference. Instead of going live at 10 a.m., they published at 09:59:59.985, giving anyone monitoring their website a 15-millisecond head start. In that time – that's *15 thousands of a second* – the high-frequency algorithms managed to squeeze in $28 million worth of trades. That's a lot of whisky.

These algorithms are not supervised by humans. They decide what to buy and sell on their own. And as a result, they can sometimes go a little bit off-piste.

One afternoon in May 2010, for no apparent reason, $1 trillion was wiped off the US stock market in the space of half an hour. The algorithms, mistaking a market fluctuation as a trend, all jumped on the same trading bandwagon and accelerated, until the Dow Jones was in free fall. Within the hour, the algorithms reacted to the new prices and reversed their trading direction until the market had almost fully recovered; but anyone who had bought or sold that day was still stung by the swing of almost a thousand points.

As the financial regulators started picking through the ashes of what became known as the Flash Crash, trying to work out what the hell had happened, it became clear that there was a gigantic obstacle in the way. Trades had been stamped with the time that they occurred, as is expected, but everyone's clocks were set to slightly different times. It was as if your convenience store whisky receipt read 11:38:00, but your credit card statement read 11:37:45, making it look as though you'd bought the bottle before it was sold. That would be easy enough to untangle in the case of one single transaction; somewhat harder when you're unpicking all the trades on the Dow Jones for an afternoon.

Worse still, the algorithms were only marking their trades with the time to the nearest second, even though any one of them could run tens of thousands of trades between one second and the next. It rendered the forensic analysis that followed an impossible task – a bit like trying to understand a football game when all you've got is an unordered list of passes between players.

You can't do it. No one could definitively pinpoint what caused the Flash Crash, and they still can't. The regulators clubbed together to set their clocks and fix the problem. All the trading companies would have to time-stamp their deals to the microsecond (that's one-millionth of a second). Everyone would have to make sure their clocks were perfectly synchronized.

And how would they sync up their clocks?

With a cable: a physical connection, hardwiring the banks all the way to the Master Timekeepers – a direct line, connecting them at the speed of light to the most accurate atomic clocks in the world.

In the UK, it's a fibre-optic cable that runs nearly 20 miles under the River Thames and the streets of the capital. The National Physical Laboratory – the modern-day Ruth Belville – sells access to the cable to the bankers.

With the cables in place, the regulators were happy.

Everything was fixed, and the financial markets were never problematic ever again.*

Bankers still hate the leap second. The physicists of the world have worked out that the only way to know what time it is is by agreeing what time it is: atomic time from some 70 clocks, occasionally knocked into shape with leap seconds so the Earth can keep track.

As yet, however, there's no agreement on when one of these seconds should be added. Some companies add it to their systems on the stroke of midnight, others an hour before. Some, like Google, spread the second out across the course of an entire day.

When you're in the business of microseconds, that discrepancy can cause quite a headache. The future of time, as a result, is not set in stone. There is a furtive debate going on about it, and leap seconds might well be abandoned altogether, making atomic time alone the ultimate arbiter.

Which brings us to the most important point. The answer to the question 'What time is it?' is actually 'What time do we all want it to be?'

'That's relativity'

There's just one more thing to consider. All of this goes out of the window once you take relativity into account.

When it comes to the scientific time lords, Einstein is

---

* Oh, except that, as well as 2010, flash crashes also happened in 2013, 2015, 2016 and 2019; and despite calls to ban the predatory practice of high-frequency trading, it remains very much legal. Sleep well!

perhaps the granddaddy of them all. His contribution to physics, astrophysics, cosmology and our understanding of the universe is arguably unrivalled. At the beginning of the twentieth century, Einstein threw everything we thought we knew about time out of the window. With his theory of Special Relativity, he worked out that time itself is not fixed but flexible – it completely depends on the position from which you are measuring it. A second is always a second, but if you are on the surface of the Earth looking at an astronaut's clock in orbit above you, then your second will pass a tiny bit more quickly than hers. And if she travelled to the edge of a black hole, then her second hand would look to you like it had stopped completely, but to her it would tick on as normal. Many books have been written about this bizarre phenomenon of time dilation, one of the most important and brain-scrambling concepts in physics. You'll be relieved to hear that this book is not one of them.

Instead, Einstein's contribution to our particular story comes in the form of a famous quotation that he may or may not have ever said, but serves our purposes rather well:*

When you sit with a nice girl for two hours you think it's only a minute, but when you sit on a hot stove for a minute you think it's two hours. That's relativity.

It's all very well knowing what time it is to the nanosecond so that you can make a shedload of cash, or know the nature of

---

* Many versions of this exist, and often when quotations are so perfect they turn out to be apocryphal. It does seem to be the case that Einstein's assistant Helen Dukas distributed this phrase to journalists from 1929, and since then it has become legend.

spacetime itself. But what really matters to most people, most of the time, is: 'How has this year gone so quickly?', 'Will this piano recital go on for ever?' or 'How much longer do I have to pretend to be listening in this meeting?'

Like all living things, we are four-dimensional creatures, passing through time as well as space: the way we *perceive* time is just as important as the real time, and from our perspective, the passing of time is very much not constant. Yes, the days vary because of our wobbly orbit. Yes, a second on Earth is infinitesimally shorter than if you're standing at the event horizon of a black hole. Here in our lived lives, the seconds tick and tock, the oven clock is still wrong, but we are the worst timekeepers of all.

The body clock

We have our own internal clock – known as our circadian rhythm. It ticks along, nudged and adjusted by daylight, but largely controlled from within our brains and cells. Lots of animals have their own versions, making some nocturnal – owls, slugs, vampires (see box over the page) – and some crepuscular – pandas, wombats, ghosts (see box) – meaning that they operate during the twilight or dawn. Humans, however, are diurnal, meaning that we operate mostly during the day and sleep at night, and our body clock has evolved to follow this pattern. The circadian rhythm is a physical process inside our bodies, regulating our blood pressure as the day unfolds (it's lowest in deep sleep around two o'clock in the morning, and surges around dawn as part of the process of waking up). It determines when we are at our sharpest (mid-morning) and our

GHOSTS AND VAMPIRES

Both of these are true – sort of. One of the more convincing medical explanations for the existence of vampires in folk tales from many cultures involves a suite of real blood diseases called porphyria. In some versions, toxic chemicals are deposited in the skin, which causes extreme light sensitivity, and patients complain of burning pain in daylight. Other symptoms include skin disfigurement, gums bleeding and receding to reveal more of the teeth, and an aversion to foods that are high in sulphur – such as garlic.

As for twilight ghosts, this is down to the way our eyes are built. The cells in our retina that turn light into sight are called photoreceptors, and there are two types. Cones detect colour and are positioned in the centre of your retina. Rods detect movement and black and white, and are located in the periphery of the retina. What this means is that when we see things moving out of the corner of our eyes, in low lighting, they tend to be devoid of colour or resolution. Combine this with a psychologically spooky situation, such as a graveyard at dusk, and your brain will go into overdrive trying to explain what you half saw, when it was probably what we like to call crepuscular witches' knickers – a plastic bag snagged in some branches.

Ghosts are almost never seen in full colour or in daylight, because the anatomy of our eyes doesn't permit it. Vampires are never seen in daylight because of porphyric photophobia. That, and the fact that vampires and ghosts don't actually exist.

most coordinated (mid-afternoon), and even tells us not to poo at night, but save it for just after waking up.*

By now, it won't surprise you to learn that, regulated as it is by light and a smorgasbord of hormones, our internal clock is not something you could set a clock to. On average – and bear in mind that there is a lot of leeway here – the diurnal cycle of a human is 24 hours *and 11 minutes*. Just a shade longer than one rotation of the Earth, then (on average, plus or minus some iron core sloshing). That doesn't mean that we keep accelerating away from perfect synchronization with the day and night, and that after just a few weeks we are hours out of sync. It just means that our bodies are constantly having to readjust in order not to get hideously out of whack. It's one of the reasons you're always tired: your biology is telling you that you should probably get off Twitter and have an early night.

Other than in extreme and very unnatural examples of isolation (see box over the page), our body clocks don't seem to have much influence on our perception of time. The fact that your body will suppress your bowels at night, and sharpen your senses for the daytime, does little to explain why lunchtime still seems an age away, or conversely how the entire morning has passed in the blink of an eye and you've only managed to type one sentence.

---

* There has been much speculation about what Donald Trump was actually doing when, during his tenure as US President, he tweeted most frequently and, by extension, produced the most typo-filled and unhinged messages. A study in 2017 accounted for this exact question by analysing 12,000 tweets by the (then sitting) President, and demonstrated that the time he most frequently pushed out his messiest effluence was indeed just after dawn, though whether he was actually a sitting President at that moment is unknown.

LIVING UNDERGROUND

Humans are social beings, and generally function best when day follows night. When those normalities are removed from our lives, our sense of time goes haywire. Solitary confinement in gaols is a cruelty applied as punishment for wayward prisoners, but its effects are better understood from wayward scientists getting rather too involved in their work. Michel Siffre, a French geologist, was leading a two-week expedition to study a glacier deep under the Alps in 1962 when he decided to stay for two months. During this time, he had no access to time other than what his body decreed. There were no clocks, no sunlight or moonshine, and he would communicate with his colleagues stationed near the entrance of the cave only when he wanted to, so they could not inadvertently give him chronology cues. In there, he read Plato, and slept and woke when he felt like it. He went in on 16 July and came out on 14 September. But he was convinced it was 20 August.

He wasn't the only one to take on this challenge. In January 1989, a cave explorer called Stefania Collini began a 130-day solitary confinement deep underground in New Mexico. She lived in a glass box designed for NASA experiments on isolation, with no external time cues, and only some frogs, grasshoppers and two mice called Giuseppe and Nicoletta to keep her company. She reported that her circadian rhythm expanded to 28 hours initially, and eventually to 48 hours. When she came out on 22 May, she guessed that it was 14 March.

There aren't many of these studies, because you have

to be nuts to try them, but in the few cases that we have of people living in solitary confinement for extended periods of time, they all begin to settle into a 48-hour circadian rhythm, and all radically underestimate how long they've been away. To this day, no one really understands why.

If you ask people to estimate a duration of, say, five seconds, you generally find they can do it pretty accurately, typically deviating from the correct amount by only 1 per cent or so. But there are certain disorders that have a major effect on that accuracy. The most significant is schizophrenia. Several studies have shown that people with schizophrenia significantly overestimate how much time has passed, with one study finding their average guess at five seconds came in at over eight seconds. Schizophrenia is very complex in terms of both behaviour and neuroscience, and it's not easy to explain at the level of brain chemistry, because it affects a lot of different bits of the brain. But it must be bewildering to live within a world of this sort: our perception of time ties cause to effect, and in schizophrenia this link starts to become uncoupled. One theory is that this leads people with schizophrenia to connect thoughts and actions in ways that seem to make no sense to others.

All perception occurs in the dark recesses of the brain, enclosed in a thick skull. The inputs into that lightless place may come in the form of light, or touch, or smell, but they have to be processed and turned into thought and experience. It's inside the human body that the true weirdness of time dilution occurs. Einstein's hot stove is a very real phenomenon.

A BRIEF HISTORY OF TIME

135

The everlasting hippo

Our brains twist time. To try to demonstrate this, in 2004 scientists sat a group of four study participants down and showed them one of the most boring PowerPoint presentations ever. On the screen flashed a series of ten circles, one after the other. Each image was identical to all the others, except one – a black circle which slowly expanded to fill the screen.

All through the slide show, the viewers were asked how long they thought each image had been displayed. Even in this first experiment, with only four participants, their answers hinted at just how weirdly flexible our mental clocks are. By now, versions of this test have been repeated on hundreds of people. The pictures that flash up on the screen are different for every experiment –

Shoe, shoe, shoe, flower, shoe, shoe, shoe
Cup, cup, cup, cup, hippo, cup, cup

– but the result is always the same. Viewers are convinced that the odd one out was on screen for significantly longer than the others, when in fact each image was shown for exactly the same amount of time. Our brains, honed as they are to detect novelty, registered the anomaly as lasting far longer than the boringly repetitive other images. Time seems to slow down when something strange happens. The researchers call this *Time's Subjective Expansion for an Expanding Oddball* – which would also be a good title for a concept album.

Novelty isn't the only factor that warps time: the promise of a reward also makes the time-dilation effect of an oddball even stronger. When offered points for assessing how long the oddball image was on screen, viewers were convinced that it endured even longer than when no reward was offered, even though, once again, all the images appeared for exactly the same length of time as the others.

There's more: your brain's ability to warp time is also affected by how full your tummy is. Psychologists Bryan Poole and Philip Gable told their volunteers that they'd be shown a series of different images that were going to be up for one of only two durations: short or long. They trained their participants to tell the difference – showing them image after image for precisely 400 milliseconds or precisely 1,600 milliseconds (that is, just under half a second or more than a second and a half) until they learned to distinguish between the two. Then they were given a new show – this time containing images of neutral objects (geometrical shapes), nice objects (pretty flowers) and highly desirable objects (lush pictures of some really, really delicious puddings). Much like their training set, the images were on the screen for 400 or 1,600 milliseconds, and all they had to do was work out which were long and which were short.

No matter how long the pictures were shown for, the test subjects were adamant that the puddings had only flashed up on the screen for the shortest time. Looking at delicious pictures of puddings actually made time pass more quickly. And the hungrier the participants were, the more likely they were to think the drool-worthy pictures were flicking by faster.

Poole and Gable's willing victims were all undergraduate psychology students. These tend to be the prey of choice for university psychologists – there are a lot of them around and they're quite easy to bribe.* Sometimes in these experiments, participants are rewarded with (much-needed) course credits, or (even more needed) money, but in the next stage of this particular experiment, the students who took part truly got their just – and actual – des(s)erts.

The students were split into two groups. The first group were shown 36 pictures of delicious puddings and told to guess how long each of them was on the screen. The second were shown 36 pictures of delicious puddings and told that afterwards they could eat them. Any of them.

The students who were promised pudding were convinced that the whole experiment went significantly faster than those who were told that they could only drool over the pictures.† It's not so much that time flies when you're having fun. It's more that time really flies if you're promised pudding.

Einstein had it pretty much right with his cheeky quotation, but perhaps we should update it, based on what we now know:

When you sit on a bench in front of a conveyor belt of boring objects punctuated by the occasional nice woman, or man, or delicious pudding, while you are really hungry for any or all three, *and* being paid for it, then time is really going to whizz by. But, if something unusual appears, like a hippo, time will seem to slow down. Sitting on stoves remains ill-advised.

* As a result, we probably know more about psychology undergraduates than any other group of people on Earth. For science and humanity's sake, we'd all better hope that they are relatively normal.
† In the end, both groups were given the puddings. Scientists are not monsters.

The 100-foot drop

If you've ever looked up from the clock while playing video games to realize that several hours have passed, you will recognize this phenomenon. Our brains speed up time when we're motivated by something that we want. And, using that phrase made popular by coders and tech dorks – it may well be that our distorted time perception circuitry is a feature, not a bug. Perhaps it persuades us to persist in hunting for food, or water, or gold coins, or princesses in towers, for a longer amount of time. Perhaps time appearing to pass more slowly would make those goals seem much less desirable. We can't be very much more precise than that, but it is plausible that this time-warping is an anti-frustration device to help us when we're looking for something that we really want or need (and let's face it, everybody needs pudding) by apparently reducing the anticipation time.

If getting unstuck from time like this is indeed a feature to help us survive rather than a weird glitch in the matrix, then you might expect the opposite function to exist as well. Does time also warp when something that you really don't want is about to happen? The evidence for this emerges when we consider something that is very close to the top of most people's most heartfelt desires: not dying.

People often report that in a traumatic moment they felt as if time had suddenly decelerated and faltered, like a movie going into slow motion, when you can see or experience much more of what is happening as it unfolds. This phenomenon is known as tachypsychia, and because it is so frequently reported anecdotally, scientists have tried to test it, understand it and come up with plausible explanations for it.

TEMPUS FUGIT

Adam writes: I was in a car crash once, not long after passing my driving test aged 17. I was stationary at a junction waiting to turn right, when a car ploughed into the back of me at about 40 mph. No one was injured, but both cars were written off. In the space of milliseconds, I have clear memories of thinking three discrete thoughts: (1) looking in the rear-view mirror, 'That car isn't going to stop in time'; (2) 'Mum is going to be really cross if I die, and she will never let my [younger] brothers drive now'; and (3) 'Oh dear Lord, that is my brains splattered on the windscreen.' Mercifully, it wasn't: I was on my way back from my gran's house with some raspberries for my sister's birthday cake, and in the impact they had hurled themselves from the passenger seat on to the glass, just to top off the drama with an unnecessary cherry (or raspberry). As for other people who have been in traumatic scenes such as this, time felt like it had slowed down because my brain seemed to have processed so much more information than normal in such a short period of time. I had calculated that the crash was impending, pondered a weird expression of guilt and the familial consequences of what was about to happen, and reflected on the possibility that I was already dead in a particularly horror-film way. My mum wasn't cross, of course, she was infinitely relieved that I wasn't injured. But the cake was indeed fruitless.

Hannah writes: When I was an undergraduate, I was walking home one night through King's Cross in central

London when I saw that there was a serious street fight about to kick off, one of the type that are best avoided by turning round and walking away, or surreptitiously hiding. I opted for the latter, and ducked into a kebab shop to get clear of the trouble. This, however, turned out to be a poor choice, as the brawl spilled over into the same kebab shop. Sadly, and terrifyingly, the fight escalated, someone pulled a knife, and a man was stabbed in front of me (fortunately it was not a serious wound, I later found out). Did time slow down for me in this moment of life-threatening violence? Unfortunately not. The reason I was walking home early was because my contact lens had fallen out and I couldn't see a bloody thing, let alone the horrific situation unfolding in front of me. It seems tachypsychia is dependent on actually being able to see.

One interpretation is that because our brains have recognized that a life-threatening situation is playing out, we are allocating many more resources to sucking in as much info as possible in order to prepare our escape or work out how to avoid death.

But it's hard to confirm this explanation, for a couple of reasons. The first is that, after the event, we may be attributing the slowing of time to the fact that we've absorbed more information than we would normally, and the best way to process all this extra info is to persuade ourselves that time itself warped. The second thing is that the question remains whether time dilation in an accident is a cause or an effect. Do we experience slow motion at such a traumatic moment in order to process more information, or *because* we are processing more information?

These are hard questions to examine, especially given that doing so would require experiments that involve real danger, posing a somewhat harder challenge to a scientific ethics committee than whether to give students free puddings. But the neuroscientist David Eagleman designed just the experiment to do exactly this. He and his team built a really bad chronometer, one that was basically a wristwatch version of an old childhood toy – the one where a paddle has a picture of a bird on one side and a picture of a cage on the other. Spin the paddle fast enough, your brain merges the images together and it looks like the bird is sitting in the cage. Eagleman's wristwatch also flicked back and forth between two images: a pixelated picture of a red number on a black background, and its inverse – the same number in black on a red background. Whenever the flicker rate is slow, it's easy to tell what number appears on the watch. But turn up the speed and your brain combines the two images, leaving you with nothing but a screen filled with a reddy-black blob and no idea what number you're staring at.

Eagleman persuaded some volunteers to strap the chronometer to their wrists, setting the flashing numbers at a rate just slightly too fast for humans to be able to read them, and threw them from a 16-storey height.

More precisely, he took them to a Dallas amusement park where one of the rides is a totally free-fall drop from 31 metres into a large net. There's no rope, no bungee cord: it's a genuine free fall, lasting two and a half seconds. The hypothesis was that if the reports of tachypsychia are correct, then in this two and a half seconds of sheer terror the subjects would be able to see the digits that otherwise were going too fast for our minds to recognize in normal, non-terrifying situations.

The results? The digits were still a blur, and none of the participants could see them. Admittedly, one woman had squeezed her eyes shut the whole way down, but even discounting this false negative, the numbers were still flashing too fast for any of the volunteers to perceive them. All the participants reported that they felt they had fallen for much longer than the two and a half seconds it had really taken. When they were asked to demonstrate how long they had been in free fall with a stopwatch, they typically overestimated by a third. Their brains were convinced that time had slowed – that they had more time to take in what was going on while they fell – but none of them seemed to have gained any supernatural powers on the way down.

These two facts seem at odds with one another, but there is an explanation. What we now think is happening is that our perception of time bends and warps depending on how hard our brains are working.

During a fall or a dramatic event, when you become hyperaware of your surroundings, your brain records more memories. Looking back, even immediately afterwards, when you reconstruct the experience in your mind, all that extra information leads your brain to conclude that the event had a longer duration than it did in reality. It's your perception of time that is warped, not time itself.

That's because, alas, our neurons don't have the power to change the laws of nature. But our minds radically change how we perceive reality. The threat of danger or excitement activates our brains to register more, and record more, of that particular situation. We construct reality inside our heads, where the mechanics and psychology of our brains give us time-bending skills.

It is tempting to speculate that this time-warping power evolved in our ancestors to equip us better to rapidly come up with a strategy to avoid danger. It's also tempting to think that this same phenomenon is not unique to humankind. It would be just as advantageous to a cat, bird or gazelle to have a super-alert mode that it can switch on in a moment of threat, and that could well mean that this time-warping ability is something originating in a deep evolutionary past. But we're only just beginning to be able to test this in humans, and we have as yet no idea how to do so in cats, birds or gazelles. Maybe, one day, one of you reading this book will come up with an experiment that will do just that, and discover that time-bending is not unique to human beings.

Navigating the flow of time

Time is a property of physics, inseparable from space. The planets and stars tick and tock with enough regularity that we can follow their movements thousands of years into the past, and millions into the future. In that sense, we live in a clockwork universe. It's not perfect, but its celestial bodies are predictably irregular. The Earth spins on its axis to give us night and day, and spins around the Sun to give us a year. Orbits bend, warped by the gravity of other bodies in space, and planets wobble because of the molten metal cores sloshing around in their guts. These celestial blips are enough to make them poor timekeepers. So we turned from the skies to the atom to nail down time, and decide what a second truly is.

But even with that precision – a clock that will only go wrong by a single second in the duration of a universe – there's one big problem with the flow of time: us. We humans are a strange feature of the clockwork universe, whose minds have the power to alter time, impelled by joy, injury, accident, isolation, novelty or even money. Our brains process the passing of time in a way that is highly dependent on what we are doing – whether it is pleasurable, boring, painful or life-threatening. To us, the passing of time is not fixed. No matter how accurate a clock we can build, our experience of time is subjective, and depends on our psychological state from moment to moment. What we feel is how we experience not just time, but light, taste, smell and every sensory input that enters the dark recesses of our brains. Experience is what colours our existence.

Your authors are scientists, and we therefore subscribe to the view that there is a very real universe, made up of physical matter and governed by rules which, at their most fundamental level, are non-negotiable.

But we are also human, and so we experience the universe in our own heads, which are places crammed full of peculiarities and biases, of behaviours evolved to help us negotiate the world, and behaviours that are strange by-products of that very same evolution. Our minds are vast spaces that absorb, interpret, process, filter and sometimes deny reality. To put it bluntly, we humans are wondrous beings, who can transcend time and space with our inventions and knowledge. And we are simultaneously deeply flawed, and absolutely atrocious at seeing this amazing universe as it really is. The first step to true enlightenment is knowing this very fact.

CHAPTER 6

LIVE FREE

This book is crammed with origin stories. Big Bangs, planets, gods and monsters, even life itself. All are tales of how we came to be what we are today. This process of understanding ourselves by looking into the past is found deeply rooted within our culture. It's the method we deploy to know and comprehend our most beloved characters. Take plucky Jane Eyre: orphaned and abused as a youngster, she grows to be fiercely independent, driven by a sense of freedom while being paradoxically prone to falling in love with handsome jerks.* You can't understand who she is unless you know where she came from. In the real world, there is the shocking but ultimately life-enhancing story of Malala

* If it is not apparent already, we are not literary critics; please direct all comments to our GCSE English teachers, Mr Kitchen and Miss Corbit.

Yousafzai, shot in the head aged 15 by Taliban terrorists in an attack provoked by her desire simply to go to school. But she recovered, powered on to redouble her passionate advocacy for women's education, and in 2014 became the youngest Nobel Laureate in history. Or there's the young boy whose parents were murdered in front of him in a dark alley, an act of mindless violence that drove him to dress up like a giant bat and psychopathically beat the crap out of unnecessarily flamboyant minor criminals. On reflection, perhaps the Malala story is a better example.

Our lives are lived forwards but understood backwards. History is contingent on the order in which things happen. The cosmic happenstance of events always decrees the outcome. A glass smashes on the floor because the cat pushed it off the table. The glass didn't smash before the cat pushed it. That would be impossible, because time's arrow only flies in one direction. Cause *must* precede effect.

This non-negotiable fact of life, however, introduces a subtle but important puzzle. When looking backwards, every effect has a cause. But when we look into the future, we don't feel as though our actions are tied to what has gone before. We feel unshackled by past events and completely free to choose our own paths. Can both things be true?

Was the cat always going to smash the glass, or could it have been prevented? Did the murder of Bruce Wayne's parents mean Batman was fated to exist? Did you choose to read this book? Were you always going to leave a glowing five-star review of it on a popular online bookshop? Are we all merely puppets on the invisible strings of cosmological forces well beyond our understanding, let alone our control?

We don't shy away from the big questions in this book, and we're in exalted company: Aristotle, Plato, Descartes, Confucius,

Sartre, Bart Simpson[*] and all the big cheeses of philosophy have tackled the question of free will over the last couple of thousand years. We also like to poke around the edges of serious subjects and prod at them sideways. So our story begins with a minor player in the history of thinking about stuff, but one who played a major role in questioning our fates.

The nineteenth-century Belgian astronomer Adolphe Quetelet was a data junkie and hoarded reams of information on the role of uncertainty in human lives. Using a national collection of crime records, he became one of the first people to apply the techniques of the hard sciences to the messy and unpredictable realm of human behaviour. What he discovered in those numbers was that we're much more predictable than anyone had predicted.

When Quetelet (who was born in Ghent, then part of the French Republic) looked through years of French crime records, he was astonished to find that they never really changed much. Regardless of what the courts and prisons did, the number of murders, rapes and robberies seemed to be the same in each and every region of France year after year after year. There is a 'terrifying exactitude', Quetelet said, 'with which crimes reproduce themselves'. Even the methods criminals chose appeared fixed: an almost identical number of murderers would decide to kill with guns, swords, knives and canes every single year. 'We know in advance', he went on to say, 'how many individuals will dirty their hands with the blood of others. How many will be forgers, how many poisoners.'

[*] *The Simpsons*, season 1, episode 2, 'Bart the Genius'. Rick and Morty might be a better example, though, of questioning the nature of free will in animated form, notably in the episode 'Auto Erotic Assimilation', but HF has already requested that this footnote be cut due to its extreme dorkery.

If Quetelet was right, then the behaviours of people are predictable and predetermined. And if that is the case, can we still cling to our belief that we have free will? This was – and is – a troubling idea, particularly in the context of crime. If, as Quetelet said, 'Man's free choice disappears [. . .] when the observations are extended over a great number of individuals,' then how can you justify punishing people when they break the law? How can you hope to improve a society by imposing moral codes and demanding good behaviour if crime is going to happen anyway?

These were uncomfortable questions without easy answers. It certainly feels as though we have free will, but how can we be sure? How can we test whether a human has agency?

Clearly, there are certain behaviours for which we do not have a choice. A newborn baby will grip like a clamp to hair, clothes or pretty much anything, especially if it thinks it's going to fall. This instinct is so innate that foetuses start practising their clenched fists at around five months into gestation. Once born, it'll instinctively suckle a nipple or teat. It has a reflex that turns its head to seek the source of delicious milk which is called rooting. It is hard to overcome our instinctive reflexes, and if you're a week-old baby, it's impossible.

These responses are not limited to babies. If one of us were to apply a sharp tap to your patellar tendon, your leg beneath the knee should involuntarily kick. If we were to click our fingers or clap right near your eyes, you would blink and then be legitimately annoyed.

All animals have these instinctive behaviours. They are called *fixed action patterns* – things they will all do without apparently having the choice not to do them. Touch a fish on its side, and it'll perceive a slight pressure difference in the water and is virtually guaranteed to bend in a C shape to wriggle away from you. If a greylag mother goose spots an egg that has rolled out of the nest, she will nurdle it back to safety with her neck and beak. That behaviour is so hardwired that all you need to do to trigger it is to place a vaguely egg-shaped object in the vicinity of her nest: a golf ball will do it, a doorknob, even a wildly un-egg-sized volleyball. Her actions are so fixed to the trigger that you can even remove the egg while she's trying to rescue it, and she'll continue rolling her head and neck in the same distinctive pattern around an imaginary egg that is no longer there. The goose must roll.

None of these things are within the animal's conscious control. They are evolved behaviours which serve to protect its well-being, or that of its babies – that's the whole point of evolution. If we put the question of human agency to the side for a moment, what's particularly interesting when you look at free will in the animal kingdom is just how often one species has evolved to exploit the lack of agency in another.

Sometimes it's a simple act of mimicry: earthworms react to the distinctive sound of moles digging through the soil by crawling to the surface, where they know their predators won't follow them. The crafty gull has worked this out and perfected a way to tap its feet on the ground to make worms wriggle to the surface and present themselves as a tasty snack that can't run away.

Some interactions, however, go beyond a simple trick. These seem rather more sinister, involving the forcible removal of agency. The prey, against its natural urge to survive, is cast under a spell and bidden to enact the will of a parasite. We shall call these 'hypnotic mind-control zombification hexes', even though literally no one else does.

Hypnotic mind-control zombification hexes

The great biologist E. O. Wilson described parasites as 'predators that eat prey in units of less than one'. These are creatures that live in or on another – feeding on their flesh or laying eggs within their guts – and in doing so damage the host, though not necessarily fatally. There are many types of parasites, but when it comes to the question of free will, there is one type that is of particular interest to us, one that involves changing the behaviour of the host to serve the parasite's wishes – hence our proposed name, hypnotic mind-control zombification hexes.

There is a beautiful wasp, for example, with a bright emerald-green body that has evolved a hypnotic and vicious power over the hapless cockroach. It is known, in a feat of impressively uncreative nomenclature, as the emerald cockroach wasp. Even though it's much smaller than its prey, the wasp will sting directly into the cockroach's brain, injecting a heady cocktail of neurotoxins that bewitches the unwitting victim. The wasp then chews off the roach's antennae and leads it back to its nest by grabbing one of the antenna stumps, like a dog on a lead. Once safe inside its victim's own home, it will lay an egg on the leg of the totally alive but utterly zombified roach which gestates for around five days. The emerald wasp larva hatches, burrows into the abdomen of the still totally alive cockroach and feasts on its guts, before emerging as a beautiful adult from a now relieved but very dead host.

Nature is grim. And it's wily too. As much as we might like to imagine our own species as being above such things, we are not totally immune from these parasitic horror stories.

For instance, the parasite *Toxoplasma gondii* can only reproduce in the gut of a cat. That presents a conundrum for

OTHER HYPNOTIC MIND-CONTROL ZOMBIFICATION HEXES

Nature is not so much cruel as utterly indifferent to suffering, and endlessly creative. Along with the emerald cockroach wasp, there are myriad examples of impressively gross parasites that cast a spell on their bewildered hosts. These are our favourites (by which we mean the ones that we find most horrifying).

The pumping disco snail: If you happen to come across an amber snail with crazy multicoloured pulsating eyes, it's not having a wild hallucinogenic time at all. It's infected with a parasitic worm called *Leucochloridium paradoxum*, which crawls into the eyestalks and then throbs and undulates, doing a passable impression of a couple of juicy caterpillars. Once in the eyestalks, the worm releases a toxin which, among other things, persuades the confused snail to crawl out of the shade and sit proudly in the open, its eyes great tasty beacons for a peckish bird. The worm reproduces in the bird's stomach and the eggs are pooped out, to be eaten by another soon-to-be-pimped (and dead) snail. This process, where an animal pretends to be another one in order to get eaten, is known as 'aggressive mimicry', which we think is something of an understatement.

The Gordian worm: There are around 350 species of

these nasty buggers, most of which prey on crickets. The worm needs to reproduce in water, but crickets are not in the slightest bit aquatic: they spend an uninfected lifetime avoiding swimming. However, while the worm grows inside the cricket's belly it releases a chemical which makes the cricket shrug off its natural hydrophobia and commit suicide by drowning. It will hurl itself into a pond or puddle, whereon the worm emerges, up to 12 inches long, in a tangled bundle – hence being named after the unpickable knot from Greek mythology. It lays eggs in the water, which are eaten by mosquito larvae, which in turn get eaten by crickets, completely unaware that they have consumed a meal that will soon hex them into going for their first and only dip.

Crab hacker barnacles: When a female larva of this barnacle finds a green crab, she will settle on its body, preferably at the base of a bristle. She will then worm a pointy tentacle into it and eventually take over all its nutrition. The crab stops moulting, which would shed the barnacle, and the crustaceous host finds itself with an appetite only for whatever the barnacle needs. The female crab stops producing eggs, her ovaries atrophy, and she becomes a lifelong nanny to the barnacle's eggs instead. Male larvae burrow into the crab crèche and fertilize the eggs, and this process goes on for as long as the crab lives – up to two years. If the larva latches on to a male crab, it castrates the male and feminizes him, with much the same outcome. This barnacle fits into a category of creatures that are described with two words that really shouldn't be seen next to each other: parasitic castrator.

Ant mind-control zombie fungus: Exactly 25 centimetres from the ground all over the Amazon rainforest there are single ants, still very much alive, but whose minds and bodies belong to the fungus *Ophiocordyceps unilateralis*. When an ant, happily anting away in its colony, gets infected with the fungus it completely loses its mind. The fungus multiplies and forms a network inside the ant's muscles and body, taking control of both. The ant will then wander out of its home and climb up a plant until it reaches a height of 25 centimetres, where it will clamp its mandible to the underside of a leaf. At that height and humidity, the fungus thrives. The ant does not. Eventually, the fungus forms a big spike which grows out of the ant's head and develops a bulb of spores at its tip. After a few days it bursts, and because this all happens above one of the ant colony trails, it showers spores all over the next round of soon-to-be-zombie ants. It's not easy to say whether animals have free will, but we are fairly sure that this is not what the ant actually wants.

this single-celled organism: how to get from the gut of one cat into the gut of another. The parasite has come up with a three-step workaround. First, *Toxoplasma* ejects itself from the cat by stowing away in its faeces. Next – we'll leave this one to your imagination – that excrement ends up within the body of a rat. Once the parasite is in the rat, it makes its way back to a cat by getting the rat into the cat. This is the step that requires some very cunning subterfuge by the parasite because, while cats like rats, that feeling is not reciprocated.

Ordinarily, rats are cautious about being in open spaces.

Put a rat into a large room and it will tend to scurry along the sides. Add cat urine somewhere in that same room and the rat – knowing that cat wee smells like trouble – will scrupulously avoid that area. Try the same experiment with a rat infected with toxoplasmosis, however, and a very different behaviour emerges. Rats will make a beeline for the urine, even if it is placed in the open space at the centre of the room. The parasite overrides the rat's natural fear of being exposed and the scent of a natural predator. The pull of the cat urine is so strong to the infected rat that it actually becomes sexually attracted to its mortal enemy: experiments have shown that the neurons triggered within the infected rodent's brain are those which fire prior to sex. We don't know if the rats would actually try to mate with the cats if they got close enough, because cats typically have other plans.

But cat lovers beware. The mind-altering powers of *Toxoplasma* are not limited to bewitched horny rats. Humans can be infected with this parasite as well, from handling cat poo or even soil or litter in which cats have pooed. It can be a dangerous infection, particularly in pregnant women, but mostly brings mild, flu-like symptoms. Most people don't even know that they're playing host to this parasite, even though it has the power to change their behaviour. However, the range of this parasitic hoodwinking is broad, and includes making men more jealous, suspicious and surly, but rendering women more gregarious, warm-hearted and fun at parties. Make of that what you will, though it should be pointed out that humans are a dead end for *Toxoplasma* because it won't reproduce in us, and the chances of our dead bodies being eaten by cats is pretty low, no matter how warm-hearted or jealous we might be at a do.

There are other diseases that can alter human behaviour, including ones that get us in the party spirit. A study in 2010 showed that people carrying a form of the influenza flu virus were more likely to socialize in the 48 hours after exposure, which might serve the virus by giving it more chance of infecting others. Unfortunately, deliberately giving people the flu just to study how much they want to party runs against pretty much every ethical guideline there is, and so the subjects in the study were tracked after they had been given an innocuous version in the form of a vaccine. They were shown to be hanging out with significantly more people, in larger groups. We don't know if this effect is actually caused by the infection – it might be that people feel more gregarious after they've been vaccinated – but equally it might be that the virus has the power to make us want to go to the pub.

Rabies, a parasite which is spread through biting, leads to aggression and abundant saliva in dogs and can have a similar effect on some humans. Even brain tumours have been linked to personality changes (see box on next page). We sincerely hope that no one reading this is affected by either. However, since toxoplasmosis and influenza are far more common, and a substantial number of people carrying them are asymptomatic, there will be some of you looking at this very sentence whose recent behaviour has been tilted ever so slightly by a pathogen without even knowing it. It's impossible to know whether that pang of grumpy jealousy or an irrepressible desire to belt out 'Eye of the Tiger' at a karaoke bar is truly an expression of your free will, or whether you are secretly acting under the command of an invisibly small but powerful parasitic hex.

It certainly *feels* like we are in control – that if you want to clench your fist, eat that third cupcake or punch yourself in the face, you are free to make that choice. But despite how intuitively real free will seems, there are also undoubtedly

HOW MUCH ARE YOU IN CHARGE OF YOUR OWN BRAIN?

Charles Whitman was a perfectly ordinary man until he started having headaches. He had studied mechanical engineering at the University of Texas, liked karate and scuba diving, and married his first proper girlfriend after five months of courtship. 'Everyone loved him,' according to one newspaper headline.

In 1965, aged 24, after he was discharged from the Marines, he started to get severe headaches. 'Tremendous and frightening' is how Whitman described them. They came with overwhelming violent urges that he couldn't understand. In the year leading up to his death, he visited at least five doctors and a psychiatrist. Then, on 31 July 1966, Charles Whitman sat down to write what would become his suicide letter:

I do not quite understand what it is that compels me to type this letter. I do not really understand myself these days. I am supposed to be an average reasonable and intelligent young man. However, lately (I cannot recall when it started) I have been a victim of many unusual and irrational thoughts. These thoughts constantly recur, and it requires a tremendous mental effort to concentrate on useful and progressive tasks. [. . .] After my death I wish that an autopsy be performed on me to see if there is any visible physical disorder. It is after much thought that I decided to kill my wife, Katy. I love her dearly and she has been as fine a wife to me as any man could ever hope to have. I cannot rationally pinpoint any specific reason for doing this.

Whitman murdered both his wife and his mother, then phoned their employers to say that they were unable to work that day. The following morning, he went to the University of Texas, climbed a tower and shot and killed 11 people with a hunting rifle, wounding 31 others. After half an hour of shooting, Whitman was finally killed by police officers.

During Whitman's autopsy, a large, cancerous tumour was found in his brain. Though it is impossible to directly attribute his murderous actions on his final day to the presence of this cancer, some psychiatric experts speculated that its presence and the pressure it put on parts of the brain that control emotions may well have had a significant effect on his behaviour. Brain injury is known to alter behaviour; indeed, a study from 2018 described 17 cases where previously normal patients went on to commit crimes after developing a brain tumour or suffering a brain injury, their crimes including arson, rape and murder.

The presence of these brain abnormalities does not exonerate people from their crimes, but it does raise a fascinating question: how much of our behaviour is under our conscious control? Many people will endure periods of mania or psychosis, when their biology and circumstances render them unable to control their actions. Even if we do not have a brain injury, some of us are better at self-discipline, some of us are more impulsive or prone to addictions or obsessions. Perhaps free will isn't a binary choice after all, but more like a spectrum.

cracks in its façade. We know there are some actions that we can't control, we know there are creatures that can override the agency of others, and it's possible that some of our own behaviours are attributable to the pathogens that we carry or to physical changes in our brain. If there are question marks over our agency at least some of the time, how can we be sure that we have free will for the rest? Can we know that we are not puppets being pulled around by unseen forces, and we just can't see the strings?

In the 1980s, a much-lauded experiment seemed to come to that conclusion. But – as with so much in the study of human behaviour – the reported revolutionary insights turned out to be not nearly as straightforward as they first appeared.

Thinking fast and early

Read this sentence, then, at a time of your choosing, flex your index finger. Done? Now, ask yourself: did the timing of your movement feel like a conscious decision?

In 1983, this seemingly simple question became the basis of a classic experiment about volition and decision-making by the psychologist Benjamin Libet. It would go on to have a profound impact on the science of consciousness and free will.

Libet sat down his subjects in front of a screen and asked them to note the moment they made the conscious decision to move their fingers. They were also rigged up to an electroencephalogram which monitored their brain activity.

Unsurprisingly, the participants believed they first made the decision to move their finger, and then did so. But what Libet saw – consistently – was that the brain showed a swell of activity up to half a second *before* the participants registered their intention to move their fingers. One common interpretation of this was that people's brains decided to make a movement before they themselves did so consciously. It appeared that our conscious minds were not in control of our actions. The order of events was not 'decide to move finger>finger moves', but 'brain gets ready to move finger>decide to move finger>finger moves'.

The consequences are deeply puzzling, even troubling. Some took Libet's findings to conclude that our thought processes aren't driving our choices, but merely reporting back on the decisions that have already been made. Others went further, asking: if we're not active participants in our own decisions, then can we have free will at all? Libet's study sparked 30 years of intense scrutiny, with big, tough questions at its heart. Are we slaving away as zombies to automatic lives that are already perfectly planned out? Is our sense of free will merely an illusion?

The endless analysis of Libet's result has often bypassed details of the experiment itself, particularly in favour of dramatic interpretations. Many of these details involve the issue of timing. It is pretty much impossible to say precisely *when* you decided to move your finger, especially at the millisecond scale that was being recorded. In order to note when you thought you made that decision, you have to focus your attention away from actually moving your finger to registering when you decided to move your finger. That's a helluva sentence (for which we apologize), but it just scrapes the surface of the complexities of trying to understand our brains and actions.

Some critics have suggested, pretty persuasively, that the

readiness potential in our brains revealed by Libet's experiment (known by the excellent German name *Bereitschaftspotential*) doesn't actually show the brain readying itself to move the finger, but something else, perhaps coincidental with the finger action of the experiment. Researchers using monkeys found that they have *Bereitschaftspotential* before the task has even been specified, suggesting that it's a general readiness rather than one relating to the specific task of finger-moving.

As things stand, the only firm conclusion we can come to is that studying brains, decision-making and free will in humans is extremely difficult. There is currently no consensus one way or the other. Experiments with brains and consciousness have not yet revealed any neurological basis for personal agency or autonomy, while neither biology nor philosophy have yet shown substantial evidence for anything other than free will.

But if we dig down much further, beyond the complex structures of the brain, there are other clues to be found. After all, humans are made of normal matter: molecules and atoms, protons, neutrons and electrons, quarks, leptons and the trinkets that populate the quantum realm. Physicists have spent centuries trying to find out how the universe ticks along, assembling and establishing the most non-negotiable rules. Though we humans have a capacity for greatness, we are nevertheless bound to the same rules as the rest of the universe, though our senses do not witness these mechanics. Here the unresolved question of personal free will ultimately maps on to a puzzle that mathematicians and physicists have been grappling with for some time: do the laws of nature accommodate chance, or human agency? Or does the universe run on tramlines, every decision already being predetermined and predictable?

The demon

Pierre-Simon Laplace (1749–1827) is one of history's greatest
mathematicians. He's often referred to as a French Isaac
Newton, although the French probably refer to Newton as
the English Laplace. Laplace was a man whose work had an
enormous impact on engineering, astronomy and mathematics
both during his own lifetime and over the centuries that
followed. He was friends with the chemist, taxman and
guillotinee Antoine Lavoisier (see Chapter 7), and together
they produced influential work on the nature of heat. He
turned his dizzying intellect to the motion of the planets,
formulated equations that explained the tides, and at one
point theorized a star with a mass so great that its gravity
would not allow even light to escape its grasp. This idea
was so advanced, so beyond anyone's understanding at the
time, that it played no part in the celestial science that would
eventually describe exactly the same phenomenon – today
we call them black holes. All the industrious thinking of his
remarkable mind led Laplace to conclude that there was no
place for chance in the universe.

In 1814, he imagined a super-intelligent omniscient being
who knew the precise location and momentum of every atom
in the universe. If, as Laplace believed, there was no such
thing as randomness, then the entire state of the universe
would be predetermined by the recent past, and whatever lay
ahead in the future would be the direct effect of how things
presently stood. It was as though the world and everything
that surrounded it was a gigantic mechanical clock of cause
and effect – ticking forwards, tick-tock-tick-tock, each click
of the cog according to the immutable mathematical laws of

physics, with no space for error, nothing left to chance. The universe, he believed, ran on rails.

Laplace granted this omniscient being superpowers. It would be able to predict everything. It could calculate any eventuality according to the fundamental non-negotiable laws of physics. It could rewind the cosmos back to the Big Bang, or fast-forward to the heat-death of the universe. It would have a complete knowledge of absolutely everything – where every atom was at any point in time, every planet, every human. The moment of your birth, the date of your death. Chance would have no meaning. 'Nothing would be uncertain,' Laplace explained, 'and the future, just like the past, would be present before its eyes.'

Laplace was not the first to concoct such grand universal mechanics. The classical scholars had also had a crack at it. Democritus – who conceived of the atom – thought chance to be a hiding place for those too idle to think. Cicero outlined Laplace's idea almost 2,000 years before he was born, writing in 44 BCE:

. . . if there could be any mortal who could observe with his mind the interconnection of all causes, nothing indeed would escape him. For he who knows the causes of things that are to be necessarily knows all the things that are going to be [. . .] For the things which are going to be do not come into existence suddenly, but the passage of time is like the unwinding of a rope, producing nothing new but unfolding what was there at first.

But it was Laplace who pinned down the idea, which became known in the twentieth century as Laplace's demon (see box on next page). Despite being a very big deal in laying the foundations of probability theory, Laplace thought that uncertainty was ultimately just down to lack of knowledge, which in turn was a product of our inability to see the world as it really is. He said that science had just invented the idea of chance and probability to make up for our fragility and the fact that we know less than his demon. Chance, he wrote, was 'but an expression of man's ignorance'.

This is not just idle brain-noodling. The question of determinism is fundamentally important to our understanding of reality, and – as is the theme of this book – our profoundly limited grasp of the universe means we are singularly ill-equipped to answer it. On top of that very human failing, this question comes with some heavy philosophical, scientific and theological implications. Historically, Christians have struggled with the notion of a fully omniscient being who knew the movement of every atom in the universe. Typically, that role went to God, but if He had already planned out the action of every person down to the last atom, this meant human free will and indeed moral responsibility were indeed an illusion.

There's a softer version of this idea too – one in which He knows what's going on in every scenario that has happened or will happen but doesn't intervene, even though he knows the outcome. Sort of like Stan Lee in the Marvel comic universe, God is ever present, always watching.* Meanwhile, C. S. Lewis – when not telling tales about wardrobes, witches and ~~Jesus~~ a lion in Narnia – was a thoughtful writer on the nature of Christianity, and he (and others) invented a get-out clause for

---

* Which implies that Stan Lee is God. We are not necessarily arguing against that position. Face front, True Believers!

LAPLACE IN SCIENCE FICTION

The nature of free will in a deterministic universe is explored in many of the great works of science fiction, but most interestingly in two of our favourites. Kurt Vonnegut's *Slaughterhouse-Five* is a wonderfully rambling yet perfectly clear morality tale of the horrors of war and the nature of fate. Billy Pilgrim, the main character, is a veteran of the bombing of Dresden by the Allies during the Second World War, and after returning home he occasionally and unexpectedly becomes 'unstuck in time', meaning that he fairly randomly gets sucked out of what he thinks is the present and is deposited in an entirely different time. This forces us to reassess his understanding of fate and free will. He's taken to the world of an alien species called the Tralfamadorians, which look like green hands with an eye in the palm sitting on top of a toilet plunger. Instead of progressing along time's arrow as we do, they only see time in its totality at all times:

The most important thing I learned on Tralfamadore was that when a person dies he only *appears* to die. He is still very much alive in the past, so it is very silly for people to cry at his funeral. All moments, past, present, and future, always have existed, always will exist . . . When a Tralfamadorian sees a corpse, all he thinks is that the dead person is in bad condition in that particular moment, but that the same person is just fine in plenty of other moments. Now, when I myself hear that somebody is dead, I simply shrug and say what the Tralfamadorians say about dead people, which is 'So it goes.'

The 2020 television series *Devs* also deals very explicitly with the concept of free will in a deterministic universe. A tech genius and CEO of a Silicon Valley mega-corporation has cracked quantum computing, increasing the power of computers by inestimable orders of magnitude. He builds a machine that is capable of modelling the movement of every atom – it is a *de facto* Laplace's demon – allowing him to visualize the history of the universe. But for reasons that we will not tell you so as not to spoil the plot, the machine can only see two months into the future. That computer, by the way, an elaborate system of golden tubes and lights, *somehow* ended up in Hannah's kitchen . . .

this conundrum by suggesting that God is outside of time, and, rather than foreseeing events, He sees them all at the same time, rather like the Tralfamadorians in *Slaughterhouse-Five*. And finally, there's the argument that if God is omniscient in Laplacian terms, then He must know his own future actions, and therefore He can't have free will Himself. Therefore, He can't be a personal being, which is pretty central to Christian theology, and *bamf*, God disappears in a puff of logic.

Much has been written – *very* much – on the philosophy and theology of free will and determinism, and, believe it or not, some of that writing is even better than that previous paragraph. The logical wrangling and bafflements that arise from Laplace and his know-it-all demon extended well into the scientific realm too. But as the sun set on the nineteenth century, Laplace's ideas remained intact. Many clung to the idea of their own free will, but determinism – the unbroken chains of cause and effect – reigned supreme in the underlying inorganic world.

For the determinists, everything that looked like it might contradict the idea of absolute predestination – as though it was random on the surface – actually fitted neatly into the idea of a clockwork universe once you dug a little deeper. Dice rolls, coin flips, roulette wheels – they aren't random at all. They're just as predictable as the path of the Moon around the Earth, as long as you've got enough data and some flashy equations. The moment the coin flips up from your thumb or the dice roll out of your palm, their fates are sealed, even if we struggle to predict them. Indeed, in the present century, scientists have built dice-throwing robots and coin-flipping machines that will give you a guaranteed result every single time. At a fundamental level, there is no luck or chance to be found there at all. Luck relies on our inability to see things as they really are. Maybe the same was true of things that looked even harder to predict – like thunderstorms and lightning strikes. We just didn't have enough information at the time.

With the explosion of data from mobile phones and internet activity since the turn of the millennium, some people have found renewed enthusiasm for Laplace's theories, fusing them with Quetelet's ideas about the utter predictability of people. Leviathan industries have sprung up, predicting what we'll want to buy, what we like to watch, even whom we will want to date. But some have pushed the claims of what can be predicted much further. Over the past decade, people have even asserted that – if only they had enough data – they could predict exactly which words to change in a Hollywood script to make the resulting film more profitable at the box office.

Some have said they can predict at birth who will go on to become a criminal. Some even have said that, given the right data, they could predict precisely where and when terrorist attacks will take place.

To say that we are sceptical about these claims is an understatement of galactic proportions. Data is the lifeblood of scientists. Like addicts, data is what we crave, and what we need in order to understand the world. The trouble is this: we know that more accurate data don't always mean better predictions. We know this is true because of two domains of science that were developed in the twentieth century which peered deeper into the nature of the universe than we ever had before. Both severely undermined Laplace's demon and shattered the cogs of a clockwork universe. They are quantum mechanics and chaos.

Chaos

One of the clearer explanations of chaos comes from world-renowned not-physicist Gwyneth Paltrow. In her 1998 not-terrible film *Sliding Doors*, Gwyn's character is unexpectedly fired from her job and heads home early. The narrative then splits into two. In scenario one, she narrowly catches her train, bumps into a nice chap, discovers her boyfriend is in bed with another woman, leaves him, falls in love with the aforementioned nice chap, finds out he's married, and then she gets hit by a van and dies. In the second version of events, she is momentarily delayed and therefore left on the train platform, the doors of the train closing in front of her. She misses her boyfriend's cheating, falls down some stairs, discovers her boyfriend's cheating and then meets the nice chap and doesn't die. That one tiny moment, the sliding doors of the train, set her on two different paths with wildly different outcomes, although we think they both suck.

Though we are fairly sure that Gwyn's interpretation of these two parallel plots was not derived from late-nineteenth-century French philosophy, that is where the roots of this idea come from. It was the mathematician Henri Poincaré who laid the foundations of chaos theory. He noticed that the paths of some objects – like a pencil balancing on its point, or the trajectory of the Moon under the Sun and Earth's gravity – were very difficult to predict. A perfectly symmetrical pencil could, in theory, balance perfectly upright, but even the tiniest shift from that equilibrium – a tiny gust of air, a small imperfection in the lead – would be enough to tilt it ever so slightly in one direction, tip it off balance and send it tumbling downwards. However much you could reasonably know about the pencil, even the tiniest gap in knowledge would mean you couldn't predict which way it would fall.

In Poincaré's examples, cause and effect remain unbroken, but more data doesn't help you understand what the future holds. Here's how he put it, writing in 1908:

It may happen that small differences in the initial conditions produce very great ones in the final phenomena. A small error in the former will produce an enormous error in the latter. Prediction becomes impossible.

Gwyneth, who would win the Oscar for Best Actress that same year (though for a different film), expressed a similar idea with equal gusto: 'I come home and catch you up to your nuts in Lady shagging Godiva!'

This idea – that the future can be exquisitely sensitive to tiny changes in the present – took hold in 1961, after the American meteorologist Edward Lorenz tried to run a simulation of the weather on his computer. To save time, he decided to start the simulation halfway through, using some numbers he'd printed from a previous run as his initial conditions. Rather than just crunching through where he'd left off, to Lorenz's bemusement the weather simulation promptly went off on a radically different path to the one it had taken before. The computer program was the same, so the second run should have tracked the first identically, and yet here it was – spiralling off into an alternative future that looked nothing like the previous version. It was just like *Sliding Doors*, but for the atmosphere, and with 100 per cent less death by vans.

Working backwards, Lorenz uncovered the subtle switch that had caused the paths to diverge. The computer's calculations were working to six decimal places, while the numbers he'd printed out were running only to three. It was like the lopsided lead in Poincaré's pencil. A tiny, seemingly insignificant difference between the two numbers, once inside the machine, snowballed to create a wildly different result, sending the simulated weather on an entirely different path to the one he'd expected.

Lorenz realized that this wasn't a special case. There were some systems – the weather, or double-jointed pendulums, or clusters of planets in the sky – where it didn't matter if you were working to ten decimal places or 10 million. If you started again with 11, or 10 million and one, you could end up with an entirely different answer. Tiny differences today had the power to lead to dramatic and unpredictable outcomes tomorrow. 'Chaos', Lorenz explained, is 'when the present determines the future, but the approximate present does not approximately determine the future.'

There's an awkward conclusion lurking somewhere in there. If the universe unfolds as it does in Gwyn's romcom, then even if it apparently runs like clockwork, the teeniest tiniest change in the mechanism – a fleck of dust on its cogs, a slight imbalance in one of its springs – will send events in a very different direction. This means that even if you know the underlying laws of nature, even if there really is no such thing as chance, even if the universe *is* a chain of cause and effect, unless you can measure and calculate everything to a literally infinitesimal level of detail – unless you can know the location, velocity and momentum of every atom in the universe – Laplace's demon is dead. Without that level of detail, prediction becomes impossible.

Laplace was wrong: chance isn't just a euphemism for human ignorance. It's a necessary part of our scientific understanding of the universe. If you have no idea what path you're on, probability is all you have to help you. It's the reason why tomorrow's weather forecast always gives you a percentage chance of rain. When it comes to Earth's atmosphere, knowing exactly what the future holds will always be just out of reach.

Yet, as tricky as chaos makes prediction, it still doesn't mean the universe is random. The clockwork mechanisms might not be Swiss-made, but we're still ticking forwards to the future, effect following cause. Thunderstorms and hurricanes don't just pop up from nowhere in the atmosphere, celestial objects don't just whizz off in random directions. Particles don't just pop into and out of existence by chance alone.

Or do they?

The quantum realm

Atoms are made of protons and electrons and neutrons. Protons and neutrons are made of quarks and gluons. Light is made of photons. Electrons, photons, quarks and gluons are thought to be fundamental particles. They're the Lego bricks of matter – you can't take them apart.

These fundamental particles sometimes, in certain scenarios, do things that are unpredictable. This is a bit vexatious, because matter at above-atomic level behaves fairly predictably. The mechanics of crotchety old apple-magnet Newton are precise enough for us to catch a ball, fly a spaceship to another planet, and calculate the trajectories of stars thousands of years into the past and millennia into the future. But once we go subatomic, mysterious things start to happen – for example, when you do something as simple as shining a beam of light at a sheet of glass. Einstein showed that light is a stream of tiny particles – photons, but direct light towards glass – like your window – and some of the light will pass through to the other side, while some will be reflected. The photons appear to be choosing their own paths away from the glass – some transmit through it, others reflect from it. As far as anyone can tell, which path any of the photons takes is a genuinely random event.

The subatomic world is beset by uncertainty. At the Large Hadron Collider in Switzerland, the particle accelerator that straddles the French–Swiss border smashes protons into one another at speeds very close to that of light. When two protons come together, the collision results in an explosive mess of different particles: they shatter into quarks and leptons, spraying themselves into beautiful patterns that physicists spend years poring over for clues to the structure of the universe. Once in a while, they'll detect a particle or an energy field that we've not seen before. That's the story of the Higgs boson – a particle predicted to stabilize what is known as the Standard Model. That is, the Higgs *had* to exist for all the other subatomic particles that we had detected to behave the way they do. But it took until 2012 for us to finally find one. The ingredients of the collisions in the Hadron Collider are the same every time (two protons), and the conditions are too (smashed into each other at high speed), but which particles emerge from any given collision is an indisputably random result dictated by chance alone.

In 1926, the German physicist Werner Heisenberg stumbled on something particularly odd: it was impossible to know both the position and speed of a particle simultaneously. To work out precisely where an electron is, for example, you need to shine a light on it. That packet of light is itself a bundle of energy, a photon, which will change the speed of the electron in an unpredictable way. The more accurately you try to measure the electron's position, the shorter the wavelength of the light must be, and the higher the energy the proton imparts on to the electron, the greater the resulting uncertainty of its speed will be. This leads to one of best physics jokes of all time:

Heisenberg is driving down the motorway when he is pulled over by a police officer.

OFFICER: 'Excuse me, sir, did you know you were going at 83 mph?'
HEISENBERG: 'Oh great. Now I'm lost.'

There's no way round this. At the quantum level, the world really is *built* from uncertainty.

These revelations were further nails in the coffin of Laplace's demon. With chaos, anything less than knowing the exact position of all atoms in the entire universe leaves you unable to predict the future. With quantum mechanics, knowing the position of these atoms is an impossibility – at least in any useful sense. Very small things and very complex things are inherently probabilistic.

Combine chaos with quantum mechanics and you end up with a totally tangled web of possible futures. A blip of randomness at the quantum scale could disrupt an air molecule, which, amplified by the grip of chaos, could cause a gust of wind that blows a branch that falls and disrupts the route you would have walked and stops you seeing the job advert that would have changed your life. Or an electron that did or didn't fire at a random moment in Gwyneth Paltrow's brain causes her to stumble on a step and miss a train, and her mind fills with worry so she never wins her Oscar, her life takes a completely different path and she never has the stupid idea to create a candle that smells like her vagina.

Uncertainty takes many forms. It is our only way to bridge the gap between what reality is doing and what information you can practically extract from it. Our universe also doesn't quite run like clockwork; at the quantum level it has true randomness at its core. That doesn't mean the end of forecasting – we can still predict with pinpoint accuracy what time the sun will rise tomorrow, we can build planes and say with certainty that they'll fly. It does mean, however – if we've understood our universe right – that in among all the neat chains of cause and effect, there's the occasional fizz of randomness.

Many worlds

Except, that is quite a big if. Because when it comes to the strangeness of quantum behaviour, things get bafflingly beyond our own experience. The quantum world may be very weird, but it is also the real world. At the nano-scale, and at that magnification, strange is normal.

Quantum information can teleport from one place to another without anything moving between. Two particles can respond to each other's movements even if they are literally thousands of light years apart. It's as if when Darth Vader revealed that he was (spoiler alert, for a 40-year-old film) Luke Skywalker's father, Luke's twin sister Leia instantly knew as well.

Or there's the Double Slit experiment. Electrons have negative charge and mass, meaning that they have a physical body which can be accounted for. But when you shine a beam of these particles through two parallel and identical slits, the pattern they make on the other side implies that they are not discrete particles but are a wave, as if they had travelled through both slits at the same time, as water would.

The conclusion drawn from this simple demonstration is that electrons can be both particles and waves, or either. Which is an idea that has a habit of slipping out of your brain if you think about it too hard. Physicists get round this boggle by describing all possible versions of the electron as being in a superposition. They all exist, simultaneously. That's fine and fun, until you try to take a photo of it. The camera can't see the superposition of all possible versions of the electron, and so it effectively becomes one or the other – a particle or a wave.

How does the electron know it's being photographed, and thus when to shapeshift into just one version of itself? It can't. Some physicists therefore regard what they see as the only sensible conclusion. They believe that the electron must continue to exist in all other possible states somewhere else in a parallel universe. This idea, the Many Worlds interpretation of quantum mechanics, essentially imagines that at the moment the photograph is taken, our physical world branches off into many others, perhaps infinitely many, which continue to exist simultaneously, splitting still further with every possible outcome of every possible quantum event that follows. The branches are there backwards in time too; not just when the photograph was taken, but at every moment in history. There are real-life, living, breathing, actually smart people who actually, actually believe this. They're physicists who can sit upright, drink a cup of coffee and hold a normal conversation, giving the impression to anyone watching that they have maintained some kind of a grip on reality.

Perhaps the way to imagine this concept is by thinking back to the four-dimensional ball in Chapter 3. A slice through a 4D ball gives you a sphere, and a slice through a sphere gives you a circle. In Many Worlds, the universe is actually an infinite dimensional fuzz of particles that exist in innumerable states simultaneously. Our reality – the existence that we know and experience: the bed we sleep in, the friends we love, the western arm of the Andromeda Galaxy – is just a projection of that reality into the four dimensions of space and time. Change the angle of that infinite dimensional space and you get a different projection – perhaps a world which is identical in every respect to this one, except that the last two words of this sentence are around switched. Or perhaps a world where you just bought an overpriced celebrity vagina candle. Or a world a billion years ago where the Earth was covered in nothing but cheese. Or a world even further back in time in which our planet never accreted from the chunks of rock and dust orbiting a young Sun. According to the Many Worlds theory, the universe isn't random after all; there is no room for chance because everything that could happen does.

These aren't necessarily imaginary worlds. The physicists who believe in Many Worlds say that all possible histories and all possible futures, every conceivable version of ourselves and many inconceivable ones, are all right in front of us, but we have no way of accessing them. But maybe, like some Christians suggested, God is sitting outside of time, turning the infinite dimensional space in Her hands, simultaneously watching all conceivable paths of existence. The possibilities are endless.

Fate or freedom?

What does all this mean for us? The universe is largely deterministic if we have enough data, but it's also not at the same time. We live at the macro-scale of cause and effect, but are made of stuff at the nano-scale, which appears to behave very differently. When we look for the quantum, the deterministic tramlined universe fades a little, only to return with a vengeance in the most surely-that-can't-be-real but also no-one-can-show-that-it-isn't theory possible.

And what, then, of biological creatures? Do they, like the universe, follow cause and effect with the occasional spin of chance? Are animals just mindless automata, rigidly reacting to cues in their environment? If this moment in time was replayed – in this world rather than any other – with every variable identical, would a creature's next movement be the same every time?

Testing this in humans is incredibly difficult, if not impossible. We are too complex, have too many confounding variables, to be able to rewind a given scenario and play it over again to see if we'd make a different choice. Yet when it comes to animals, where replaying scenarios in lab conditions is a little more plausible, you do not always see the same outcomes every time. That might not be surprising: being perfectly predictable is not a safe strategy, especially if you don't want to get eaten. At the beginning of this chapter, we mentioned fish bending into a characteristic C shape when they sense any pressure differences in the water that might indicate something is looming towards them. This is a predictability that can be exploited. The tentacled snake *Erpeton tentaculatum* has evolved a way to tickle the water near a fish just enough to cause it to think that it has been touched and reflexively bend away from the perceived threat. The snake doesn't aim its jaws where the fish is; it positions them where the fish will be once its fixed action response has kicked in.

The cockroach, however, a true master of survival, has evolved a way to avoid predictability. It has two little appendages sticking out of its backside that are incredibly sensitive to small changes in air pressure. Try to sneak up on a cockroach and it will invariably bolt away from you, but, unlike the fish, the angle it chooses for its escape appears to be random. It's impossible to predict which way it will run, rendering it safe from a crafty predator.

The fruit fly is beloved of experimental biologists, for

reasons too many to go into here. But one that is sometimes overlooked is that we can glue them to a stick with impunity. Fruit flies exhibit a range of complex behaviours to do with their survival, all the things you'd expect of an insect: foraging, mate choice . . . erm, actually that's about it.

They make decisions involved in choosing a mate or calculating where to find food, but given the relative simplicity of the fly's mind, it is not too much of an exaggeration to suppose that they run along fairly well-defined tramlines: respond to the environmental cues with genetically encoded programmes. Outcome: fly eats/mates = happy fly.

And yet fruit flies too demonstrate unpredictability. Experiments in the last few years with tethered flies – that is, glued to a stick – have tantalized those interested in the search for a neurological basis for free choice. If you tether a fruit fly to the inside of a white drum, as one team of scientists did in 2007, you can create a kind of sensory deprivation chamber for it. Without any visual or air-pressure cues, the fly can't get any sense of where it is or where it's going, but you can persuade it that it's flying by moving the scenery, like in a pretty standard school play. There's a torque meter attached to the fly's head, a device that can deduce the twist of its body and angle of its wings, and thus determine where the fly thinks it's going. In that environment, it's impossible to forecast how the fly will fly. Its movement is not uniformly random, either. If its attempts at zigging and zagging were as though it were flipping a coin to choose its direction, then after a while there would be an equal chance that it would go this way or that. Instead, the fly seems to be alternating between tiny movements, as though exploring the local surroundings, and the occasional big hop, as though heading off somewhere new. We shy away from using the word 'choice' here, because we can't know if it is actually thinking: 'I went that way last time, screw it, I'm going this way today – that will really confuse this scientist who, incidentally, has actually glued me to a stick.'

But we can know that the fly's behaviour is unpredictable, just as we know the cockroach runs away unpredictably too. Some scientists have suggested that these rudimentary unpredictable behaviours – not quite choices, but not quite programmed responses to the environment – might be the biological basis of free will: a molecular neurological McGuffin that unshackles the creature from the script. Could these behaviours be the simplest version of what we feel in our lived lives, that we always have choice about what we will do next?

We're not so sure. Maybe the fly's behaviour is determined by a bit of apparent randomness that is actually a biological coin flip at the molecular level. Add a bit of chaos magic to boost the initial effect, and a tiny fluctuation in a neurological pathway gets amplified to eventually produce an unpredictable outcome. The truth is that we don't know. How do you discern behaviours that are programmed from those that are based on choice? And furthermore, you can have free will and still be predictable, and your decisions could be built on randomness and still not be truly free.

Whatever free will is, or isn't, there is an underlying molecular mechanism that is happening in our neural circuits. Perhaps that creates a forgivable illusion, a lie we tell ourselves to convince us that we are in the driving seat, not puppets whose invisible nano-strings are beyond our control. Maybe free will is real, and we have complete control of our destinies. Or perhaps free will is real, and we are in the driving seat, but, with parasites, randomness and chaos along for the ride, we don't always have our hands on the wheel.

Perhaps it doesn't matter. Some argue that we believe we've got free will, so our decisions are predicated on this belief, and if we found out we were wrong, precisely nothing would change. Others think that, given our ability to predict murders and other crimes with alarming accuracy, as Quetelet did – or nowadays the number of women who will choose to have a baby each year, the number of plane crashes, suicides, or people who will visit A&E on a Friday evening – we should structure society with these factors in mind.

In the end, we have little choice but to lean on Laplace's idea of chance and probability as a substitute for ignorance. It's as though, when each of us wakes up every morning, there's a small chance we will become a murderer that day, cause a car accident, get struck by lightning. Or maybe there isn't. Either way, the world we're living in is indistinguishable from one in which there was.

As for us, we didn't promise that we would answer this question, but we were always fated to ask. We know you believe that you have free will. We do too. But what we believe and what is true are often two very different things.

CHAPTER 7

THE MAGIC ORCHID

The end of the world is definitely coming. Mercifully, it's unlikely to be anytime soon, so cool your jets. Apocalyptic predictions about the end of the world, however, are very much here already. They have been made in every culture throughout history, and all have one thing in common: exactly none of them has been fulfilled.

In general, the doomsday cults tend to be super-judgey – a bit like Santa Claus deciding who is naughty and who is nice, but with somewhat harsher consequences than a lump of coal in your stocking. There's often rather a lot of horrifying violence, and the world gets destroyed, but if you're lucky your soul will be preserved in perfect and ultimate heavenly nirvana. (Or you are sent to eternal damnation in a fiery lake. It's one or the other.) The End.

Even top physicists have had a go at prophesying the end of everything in the past – and they too got it wrong (otherwise we wouldn't be here). Isaac Newton himself spent a great deal of time hunting for hidden messages within the Bible, and came to the conclusion that the doomsday clock wouldn't tick to its final conclusion until at least 2060. Here's how he put it:

> **This I mention not to assert when the time of the end shall be, but to put a stop to the rash conjectures of fanciful men who are frequently predicting the time of the end, and by doing so bring the sacred prophecies into discredit as often as their predictions fail.**

There's Newton once again proving his genius: if you're going to create a doomsday cult, don't set the date for your end of days to be within a few years of the present – make sure you're well dead before the apocalypse fails to transpire.

Unfortunately, this is not a lesson that has been well learned by some of us mere mortals. According to a 2012 Ipsos survey of 16,262 adults in 21 countries, one in seven people think the world will end during their own lifetime. It's almost as if they can't quite see beyond their own lives.

But there is a wonderfully practical way to explore what happens when people are proven wrong. All you have to do is find a cult that has set a specific date for the end of the world and see how they react when that day passes apocalypse-free. It's happened several times in the past few decades, and on each occasion psychologists have taken their chance to interview cult members who had truly believed that the end was nigh, after they discovered it wasn't. What they've found applies to all of us: that we humans are outstandingly good at defending our deeply held convictions, even when they're proved to be spectacularly wrong.

After the end of the world

The first study of this very particular phenomenon was recorded by a group of psychologists from the University of Minnesota in a book called *When Prophecy Fails*. Leon Festinger, Henry Riecken and Stanley Schachter were three professionals who clearly took their fieldwork very seriously.

In 1956 they joined a cult. The Seekers were founded by Chicago homemaker Dorothy Martin. She had been involved in the early Dianetics movement with the science-fiction writer L. Ron Hubbard, who would later go on to found the space alien celebrity religion Scientology. Dorothy had experimented with automatic writing, where participants enter a trance-like state and scribble down words that are apparently 'coming through' from someone (or somewhere, or something) else. The details are sketchy. She said she'd received messages from aliens on the (totally made-up) planet Clarion, who informed her that she was the latest embodiment of Jesus: this was clearly a kind of fusion religion that drew from both science fiction and Christianity. She had also received news that a great flood was coming to destroy the Earth, but that aliens would arrive in a flying saucer and rescue her and her followers from the end of days. This would happen at precisely midnight on 21 December 1956. The deluge would follow a few hours later.

As the clock approached midnight on that 21 December, Festinger, Riecken and Schachter sat with Dorothy's group, many of whom had given up jobs and possessions, and severed ties with family members, to be among the saved. The press had mocked the cult, thus fuelling the Seekers' distrust of authority and reinforcing their belief that they had privileged, secret information from the aliens. There they sat, waiting for the moment of truth. They had removed all metal items – zips, jewellery, even bra underwires – from their clothes and bodies in anticipation of boarding the spaceship.

The clock struck midnight.

Nothing happened. No rapture, no flying saucer. The Seekers sat and waited, and nothing continued to happen. It was not The End. It was an ordinary Saturday morning. At 4 a.m., Dorothy Martin began to cry.

What would happen next? The psychologists had assumed that the participants might abandon this absurd spectacle and admit their information was faulty.

But no. They all doubled down. The Seekers went back to their predictions and re-examined them, and lo and behold, three-quarters of an hour later Dorothy Martin received another message via automatic writing: 'The little group, sitting all night long, had spread so much light that God had saved the world from destruction.'

Rejoice! They had not only survived, they *alone* had rescued the Earth from an apocalyptic flood. The very next day, they sought the press publicity that they had previously shunned and celebrated their conviction.

There are two possible explanations here. Either Dorothy Martin was right all along, or – marginally more likely – she was not. The key point is that the Seekers had gone in too deep, had gone too far down their road, to give it all up in the face of a no-show by the Clarion saviours. The team of psychologists concluded that the cult had too much to lose to abandon everything they'd taken to be true. And so their belief was not only preserved, it was enhanced. The Seekers' full commitment to a busted prediction had been validated socially, and the more people knew about it, surely the more true it must be.

Other doomsday cults have followed very similar patterns, with similar post-(non-)apocalyptic rationalization. In 2011, a high-profile American Christian radio host called Harold Camping survived his predicted rapture date of 21 May. Followers attempted all sorts of explanations of why it had not transpired. First, it was going to be three days later, as in the length of time before Jesus rose from the dead. Three days passed: nothing happened. Maybe it was not three days, but seven – another holy number. Or maybe 40, as in Noah's flood. All those dates passed too, and there was no rapture and no flood.

So a new explanation arose. It was a warning by God, a test of their resolve: everyone would mock them when the day came and went, and only the true believers would remain faithful.

A number of Camping's followers were interviewed a year later by a journalist called Tom Bartlett. Some of them did concede they had been involved in cultish behaviour. Others edited their own past statements, many of which had been recorded online and in interviews, in order to qualify what had actually happened without accepting that they were simply mistaken or misled.

It is easy to mock such convictions, such post hoc rationalization, and many did at the time. But others in Camping's community also recognized that a lot of his followers were vulnerable and misled, that their distorted beliefs reflected an aspect of the human condition, and that compassionate

THE END OF EVERYTHING

Ultimately, the universe will end when it succumbs to the Second Law of Thermodynamics. All closed systems tend towards equilibrium, the point at which free energy is used up – a cup of tea cools down to room temperature, a ball rolls down a hill and stops. As this happens, the entropy of the system increases, and the non-negotiable Second Law says that entropy only ever increases. If we take the whole universe as a closed system, then one day there will be no more free energy available, entropy will be at its maximum everywhere, the temperature of the whole universe will become absolute zero, and it will reach a state of final and perfect equilibrium.

Fortunately, none of us will be around to witness that. The heat death of the universe, as it is cheerily known, is predicted to occur in roughly 10^{100} years' time (the universe is currently around 1.3×10^9 years old, so quite a way to go). But the Earth will face its own apocalyptic nightmare well before that.

The future is bright. A bit too bright, in fact. The Sun is getting slowly hotter, and expanding, and that means it's expanding towards us. At some point, maybe in a billion years or so – assuming we haven't already wiped ourselves out by then – it will get so hot that all plant life and much animal life will not be able to survive. Soon after that, the oceans will evaporate. At this point we, and all life on Earth, are truly screwed. By about 3 billion years from now, the planet's surface temperature will be around 150° Celsius. But the real fry-up will happen in 5–7 billion years' time when the Sun that we so love, that vital provider of all life as we know it, will become the harbinger of ultimate doom. It will run out of its fuel (hydrogen), and all the helium that is the waste product of the fusion reaction that has been burning away since the Sun formed will be so dense and heavy that our very own star will collapse under its own gravity before heating up and expanding into a red giant, three thousand times brighter than it is now. The edge will extend 20 million miles beyond the Earth's current orbit. Our pale blue dot will be engulfed in a hot red nuclear fire. The End.

sympathy was a better response than contempt or amusement. As Bartlett concluded, 'You don't have to be nuts to believe something crazy.'

The phenomenon itself is known as *belief perseverance* – clinging on to an idea even in the face of overwhelming evidence that it cannot possibly be true. This is a common trope in conspiracy theories. For people who are invested in a conspiracy, there is effectively a profound emotional and sometimes financial sunk cost in their adherence to it, and it weighs heavily against choosing to pull out. Many people prone to belief in conspiracies are distrustful of authority, and, much like the Seekers, believe that they have access to forbidden or secret knowledge. Giving up these convictions exacts a profound psychological cost, such that it may be easier to stick rather than twist.

Perhaps some of these people are atypical, and perhaps

some have psychological vulnerabilities that made them likely to join a cult, or buy into a fairly implausible prophecy in the first place. But the same pattern of behaviour isn't confined to apocalyptic doomsdayers – you can see it in everyone.

In the 1980s, researchers set out to see if they could trigger belief perseverance in a group of unsuspecting subjects. They showed a group of volunteers a series of seven sums – straightforward stuff, like 'What is 252×1.2?' – and then gave them each a calculator to check their answers. Little did they know, but the calculators were rigged. On the first round of sums, they spat out answers 10 per cent bigger than the correct ones, and then in subsequent questions got progressively wronger, so that by the seventh question the calculator was out by 50 per cent.

Some spotted the calculator was a dud right away. Others were puzzled by the answers it gave, but persevered before asking if there was something wrong with it. But about a third of the participants went through all seven problems without ever asking if something was awry. When questioned about what had just happened, a typical comment was: 'It doesn't look right, but if that's what the calculator says, then it's probably right.'

Perhaps our favourite example of belief perseverance, though, involves a rather mean ruse played on a group of 19 scientists. Even clever people, it seems, are not impervious to the tricks our minds play on us. The scientists were told that they were helping to evaluate a new maths textbook for high-school kids. In fact, they were the unwitting subjects in an experiment, which went like this. First, they were handed a cylinder and given an unusual (but correct) equation to calculate its volume. For a bit of reassurance, they could check their results by filling it with water. Then the real experiment began. Each cylinder was swapped for a sphere, and they were given a completely bogus formula that told them that the spheres were 50 per cent bigger than they actually were. When it came to checking their calculations by again filling the spheres with water, everyone immediately saw there was a problem. Remember that these were all researchers or professors at major universities. They had actual doctorates in science. And yet, rather than questioning the formula, they came up with some extraordinarily elaborate explanations as to why the experimental results didn't match the numbers derived from the formula.

There was doubt, there was discomfort and there were ad-hoc explanations, but still 18 of the 19 participants clung to the incorrect formula, rather than adjusting their beliefs on the basis of the irrefutable evidence in front of them.

Belief perseverance might seem like a baffling phenomenon, but emotional investment in an argument or a belief is a powerful thing. Our minds are not necessarily predisposed to abandoning that investment without a struggle. The business magnate and megazillionaire Warren Buffett puts it all rather well: 'What the human being is best at doing is interpreting all new information so that their prior conclusions remain intact.'

But we're not just bad at interpreting new information when it conflicts with our intuition. We're also unavoidably bad at deciding what new information is relevant in the first place. This phenomenon is known as confirmation bias.

THE CONFIRMATION BIASES OF RUTHERFORD AND FRY

Adam writes: I became convinced that I could kill Hollywood actors with only the power of my mind. I can chat about a film, or simply watch it, and within a few days one of the stars will be dead. In 2008, I bought *Brokeback Mountain* on DVD. The very next day, I read that one of the lead actors, Heath Ledger, had died. Later that same year, I was discussing the classic *Jaws* with a friend. Twenty-four hours later, the lead actor, Roy Scheider, was dead. In 2016, a few days after a viewing of *Star Wars: The Force Awakens*, Carrie Fisher, the actor who played Princess/General Leia – and my first true love – had become one with the Force.

What a terrible burden this power is! And with great power comes great responsibility, so I must constantly restrain myself, and heaven forfend that I should ever mention Cate Blanchett again.

Obviously – and fortunately – it's not true, just startling confirmation bias. The truth is I am a massive film bore, as Hannah will attest. I talk about and quote movies all the time, mostly while people are not listening. Mostly. I talk about *Jaws* a *lot*, because it's damn near perfect. There are quotes from films littered across this book, which Hannah hasn't noticed – I put them there purely for my own amusement. It's simply that I don't recall all the times when I have mentioned an actor and they haven't died the next day, because that happens every day. It is only the coincidences that I recall because they are striking.

Less explicable is the fact that I seem to look at the

clock at exactly 11.38 a.m. every day. Maybe it's because I might be thinking about lunch, but it's a bit too early to eat. I have tried hard to resolve this, by counting other times I look at the clock and noting if it's ever the same time repeatedly. But it isn't, and I am now convinced this number is speaking to me in some cosmic way. It is even more peculiar because 1138 is a secret number that features in multiple *Star Wars* films (as a prison cell-block number, on a label on Leia's helmet, as a droid's ID and in many other places). I can only assume the Force is with me. Always.

Hannah writes: There is an orchid in my kitchen which – I am absolutely convinced – is the secret source of all of my powers.

I am aware of how ridiculous this sounds. I know I'm supposed to be scientifically minded. I know I'm supposed to value logic and rationality above all else, but when it comes to this magical pot plant, I can't help myself. Even though I know it's impossible, I still (sort of) believe that this plant only flowers when things are going exceptionally well.

My husband bought me the orchid as a good-luck gift on the day of my PhD viva. It had flowers on it. I aced the viva. It flowered in 2014 when I gave a TED talk on the mathematics of love. It flowered in 2015 when I got a big promotion at university, and in 2016 when *The Curious Cases of Rutherford and Fry* was commissioned. Coincidence? It's not a coincidence.

I should add that I've killed every other plant I've ever owned. Including other orchids which were bought to keep this one company, and would sit on the same windowsill, get the same amount of light and water. They all died – but this one has been going strong for almost a decade.

There have been ups and downs in my life, of course, and the magic orchid has tracked these personal events, sometimes showing clear signs of distress. There was the time it accidentally got scorched by an overenthusiastic scented candle. A big project I'd been working on for a while fell through at the same time. On more than one occasion the leaves have turned yellow from overwatering. Any time my magic orchid starts to look like it's on the way out, my rational mind goes out the window, and I rush to its rescue – knowing that if I let the life within it dwindle, that will surely mean the end of my career.

We know what we like, and we like what we know

You'll know all about confirmation bias if you've ever had the uncanny experience of thinking about someone seconds before they phone you. It seems miraculous; but there's nothing spooky going on. Your brain is playing a trick on you, readily discarding the millions of times you were thinking about someone and they didn't call, or when you weren't thinking about them and they did call – why would you remember those? They're just normal. But when the coincidence happens, our brains tell us that something magical has transpired. No one is immune from confirmation bias. It's instinctive, and almost impossible to combat, for anyone. Including your authors, as the section on p.192 shows.

Confirmation bias is a trapdoor in our minds that makes us humans ripe for exploitation, and the consequences are not limited to doomed film stars and mystic bollockworts.*

In the current technological era we are more vulnerable to manipulation via confirmation bias than ever. Take the YouTube algorithm. It recommends videos to you on the basis of those you've already watched by analysing the viewing habits of people who clicked on the same video. It predicts that you are more likely to enjoy their favourite content than a completely random selection of videos, which turns out to be a fair assumption. The trouble is, there are concerns that it can have the effect of plunging the viewer into a confirmation bias odyssey. If you watched a video about aliens visiting rural America and intrusively probing local farmhands, then you are fairly likely to be interested in other bonkers conspiracy theories, such as that the Earth is flat, and that vaccines cause autism. Before long, you may find yourself presented with videos of people telling you that school shootings in the US were faked, and that the 9/11 terrorist attacks on the World Trade Center towers were perpetrated by the US government. People who believe such things are more likely to distrust the government anyway, and to be on the political right. In the run-up to the US presidential election in 2016, virulently anti-Hillary Clinton videos were viewed six times more than anti-Donald Trump videos.

The business model that underlies this algorithm is politically neutral – the idea is to publish content that makes the site 'sticky', meaning that people stay on it longer, so that

* The Middle English name for some orchids. 'Orchid' itself means 'testicle', as the root of this beautiful plant somewhat resembles one. As scholars, your authors are keen to bring this important information back into the public domain in the service of education.

adverts can get served and watched, which brings in more revenue. It therefore favours videos that tend to keep people watching for longer, leaving comments as they go, and these often have sensational themes. But it also means that if you have any kind of political bias in your own viewing habits – and frankly, it would be near-impossible not to – you will be served videos that not only echo that political view, but offer ever more exaggerated versions of it. And the concern is that this can have the effect of pushing people deeper into their pre-existing mindsets.

Facebook is fundamentally similar, and what you view and 'like' will steer your newsfeed ship in a particular direction. You can try this experimentally, by 'liking' things you don't actually like and seeing what happens to your feed. (Don't do this if you really value Facebook, as it can take you to some places you might not enjoy and leave you there.) In 2014, the journalist Mat Honan tried clicking the 'like' button for literally everything on his Facebook feed for 48 hours, to see what would happen. This included an article about the Gaza Strip. Overnight, his feed began to be more and more polarized, soon showing him increasingly right-wing and anti-immigration content, alongside ever more left-leaning sites, interspersed with the occasional 'A cloud that looks like a penis' and 'Stop what you're doing and look at this baby that looks exactly like Jay-Z.'

On Twitter, we tend to follow people who share our political beliefs and interests. We've always done this, of course, by reading newspapers and watching television programmes that reflect our preferences; but modern technology has made that reinforcement so much more efficient. In online social media we incubate ourselves in confirmation bias echo chambers, where the views we already hold are entrenched by exposure to more of the same.

YouTube rabbit holes and Twitter filter bubbles are a strange modern consequence of what happens when you build a system that exploits and exacerbates our psychological biases. But it's not the only industry that has sprung up around the peculiar failings of human intuition. When you're willing to tap into the flaws in our intuition, there are far stranger things on offer.

Paranormal activity

Psychics rely on confirmation bias – it is the fundamental human weakness that allows them to peer into people's minds. The first stage of giving a psychic reading is to make some educated guesses about the person you're talking to. If that person is sitting in front of you, you can estimate their age, and make inferences about their income, their social background (based on things like their accent and clothing), their marital status (are they wearing a wedding ring?) and a million other things that are visible to everyone, if you take the time to look.

The psychic can then make some deductions about what might be troubling the client by putting bland, generic questions or suggestions to them that are likely to trigger a personal response. People often visit psychics when they are troubled, and particularly when illness or death has affected them or their family. They do it voluntarily, too, meaning that they are likely to be cooperative, and therefore eager to hear messages that they can fit to their own situation and willing to filter out the chaff that they can't. We all do half the work in these scenarios, ignoring statements that are off the mark and latching on to the things that confirm what we were hoping to hear. If you are in your fifties and you are sitting in front of a psychic, there's a very reasonable chance that one of your parents has recently died, so the psychic can say something generic like: 'Have you suffered a recent bereavement? A parent?' And if it's not a parent, then the next most likely answer is a friend.

A skilled psychic can go much further. The Office for National Statistics has data on the frequency of names given to children for every decade since 1904. The most popular names for British boys in the 1930s and 1940s were John, David and William. One of the tricks psychics use is to guess the first name of someone the client is thinking of. A 50-something's parent is likely to be in their seventies or eighties, so again you can make some educated guesses. You'd be daft to say: 'Did your dad's name begin with a Z or an X?' because Zachariah and Xenophon were not in the top 100 names, ever. But if you said: 'I'm sensing a J, or a D,' then statistically you're sensibly hedging your bets. 'Is there someone special in your life whose name begins with J, and you want to tell them something special?' People who have seen psychics often don't recall that the medium was scratching around with vague questions or guesses about the names of relatives, only that they got it right.

We can do something similar right now, because we asked our editor for some data on who you might be. Around 60 per cent of people who buy science books are male. They tend to be younger than the average general reader. Around a quarter of books like this one are purchased by people aged 13–24. People who like science tend to be keen readers as well, buying around ten books a year. They are also far more charming and better-looking than average. We have a rough idea, too, of who likes our previous works, though Adam's readers tend to be a bit older than Hannah's. So there is an increased probability that you are in one or more of these brackets. You might be a teenager, reading this book because you are into science or maths, and one of your parents bought it for you as a Christmas present (it was published in October, so a useful gift which is both fun and educational – *parents take note*). If you are a teen, then we can make some more assumptions: you are worried about your friend group and your appearance. You're interested in a girl or a boy, or both, but your interest is unreciprocated, and frankly, sex is a bewildering minefield. You've definitely got too much homework, and you'd rather stay up late on social media or gaming. Some of your teachers are jerks. You've got to do a ton of work to get your grades, and you worry about exams. And your parents are always on your back about something.

Or maybe you are the parent of a teen, in which case you are probably anxious about your daughter's or son's grades, about their staying out too late, going on social media, and getting involved with some awful boy or girl. You are probably worried about your weight or not doing enough exercise. You're mildly concerned about your alcohol intake, but not enough to do anything about it, and frankly, sex is a bewildering minefield. Everything aches a bit more than it used to, and you're perpetually tired. If you're in your forties or older there's a high probability that one of your parents is either ill or dead already, and this doesn't help your encroaching fear of your own mortality. You like watching *The Crown* and are slightly embarrassed about that, but not about loving *Fleabag*, which you think is the greatest television show ever. God, you miss lie-ins at the weekend.

Or perhaps you're in your twenties and bought this book for yourself – in which case, enjoy those lie-ins: the clock is ticking.

It's fairly easy to play tricks that prey upon our very natural psychological biases. Our intuition is screaming out that it must be true. But our intuition lets us down. We tend to ignore things we don't agree with, and focus on the things that reinforce what's already in our minds. Don't believe us? Try this test.

Are we all suckers?

Please read the following statements and consider how much they apply to you (you can tick the ones that do in the handy circles):

You have a great need for other people to like and admire you. ◯

You have a tendency to be critical of yourself. ◯

Disciplined and self-controlled outside, you tend to be worrisome and insecure inside. ◯

You pride yourself on being an independent thinker and do not accept others' statements without satisfactory proof. ◯

You have found it unwise to be too frank in revealing yourself to others. ◯

At times you are extroverted, affable, sociable, while at other times you are introverted, wary, reserved. ◯

Some of your aspirations tend to be pretty unrealistic. ○

*At times, you have serious doubts as to whether you
have made the right decision or done the right thing.* ○

How did you do? If you feel that any, or indeed all, of
these statements describe you accurately, don't fret: here is a
comforting story.

In 1948, the American psychologist Bertram Forer offered
each of his students a free personality consultation. He gave 39
of them a form to fill out called the Diagnostic Interest Bank,
which asked questions about their hobbies, their hopes and
ambitions, and their personal concerns. The test was designed
to reveal aspects of their personalities, their worries and their
temperaments. A week later he privately gave each student an
individual, printed, personalized assessment, headed with his
or her name, based on the form they had completed. He asked
the students to read the assessments and, in order to validate
his test, score them for him according to how accurate they
thought they were (on a scale of zero to five, zero being 'poor'
and five being 'perfect'). Each personal assessment contained
13 sentences, and the ones listed above are samples taken
directly from Forer's original assessments.

When he compiled the scores, he found the average
given by the students was 4.3 – pretty much all of them
thought that Forer's assessment was a breathtakingly accurate
picture of their personalities. What else would you expect
from a top psychologist?

But then, as if from a magician, came the reveal.

In the lecture theatre, Forer got a student to read out one
of his personal statements, and then asked the class to put up

their hands if theirs were at all similar. Every hand shot up. The same happened for the next one, and the next one, until the lecture theatre erupted in laughter, when they all realized they'd been conned. In the words of the scientific paper in which Forer wrote up his methods and results, entitled 'The Fallacy of Personal Validation': 'The data clearly showed that the group had been *gulled*.'

Every single report was identical to every other. Each student had read these sentences and believed that they applied personally to her or him; but in fact they were all word-for-word the same. Even more damningly, Forer later revealed that all the sentences that had felt so insightful, so personal to each student – and maybe to you too – had been pinched directly from an astrology magazine that he had bought from a street news-stand.

Bert Forer had uncovered that fundamental weakness in how our minds work. The human psyche, fragile and complex as it is, is prone to seeking out things that confirm what we already think, and ignoring things that challenge its preconceptions. The blandness of those statements wasn't enough to prevent most people finding them both highly insightful and deeply personally relevant to themselves. In fact, they're just normal human feelings that most of us experience. Who doesn't want to be liked? Many people sometimes feel bold and extroverted and at other times just want to be left alone. Who isn't insecure about something?

These types of phrases have become known as Barnum Statements, after the circus impresario P. T. Barnum, famous for the line, 'There's a sucker born every minute' (though he didn't actually say it: see box opposite). They are the basis of all astrology, and a whole suite of other supposedly paranormal occurrences.

THE GREATEST SHOWMAN

You might know P. T. Barnum as portrayed by Hugh
Jackman in the rather good musical *The Greatest Showman*.
Born in 1810, Barnum was indeed a great showman, and an
equally great snake-oil salesman, confidence trickster and
fraud – but, as far as we know, not prone to bursting into
song and dance. Barnum's shtick was to make shedloads of
cash from promoting hoaxes and fakes as scientific wonders
of the world, presented alongside his various circuses and
museums. However, that famous 'sucker' line was not said
by him at all.

The story begins with archaeologists digging on a
farm in Cardiff, upstate New York, in 1869. There, they
uncovered the petrified body of a 10-foot-tall man, which
they presumed to be the fossilized remains of an ancient
Native American. Little did they know that the body was
not a petrified giant, but a statue (and not a very good
one, by most accounts) that had been planted there a year
before by George Hull, the cousin of the farm owner, who
had suggested that very spot for the archaeologists to dig.
News quickly spread of this amazing discovery. Church
leaders reverently verified it, pointing out that the Bible
indeed does refer to giants,* and people queued and paid
to view the Cardiff Giant in his tomb in hushed veneration.
But it's not easy to keep such a hoax secret – or to cover
your tracks when you've commissioned a 10-foot statue
and buried it on a farm. So Hull acted swiftly, and sold the
now-excavated figure to a syndicate of businessmen led by
one David Hannum for the princely sum of $23,000.

They took the giant on a touring show all around New York State.

Enter P. T. Barnum! Never one to miss out on a big scam, he offered Hannum $50,000 for the Cardiff Giant. Hannum said no. So what did Barnum do? He made one of his own. He sent a minion to sneak into the show, secretly take some measurements and carve a mini-model out of wax. Barnum used this to make a replica, which he then displayed in a museum in New York, raking in the cash from even more gullible punters who believed it was the real fake. On hearing that his scam had been scammed, it was David Hannum who uttered the immortal line: 'There's a sucker born every minute.'

Wonders of nature were all the rage in those days, and Barnum had already made a killing with his highly successful show 'Barnum's Grand Scientific and Musical Theater'. One of the stars was Tom Thumb – a cigar-smoking 4-year-old boy billed by Barnum as a minuscule 11-year-old. And there was the famous Feejee Mermaid.

Summon the image of a mermaid in your mind's eye. Maybe you are thinking of a beautiful maiden with long hair, a shell bra and an elegant fish tail? Or maybe flaxen-haired Ariel from Disney's *The Little Mermaid*?

Nope. Nope nope nope. Replace that image with the desiccated remains of the top half of a dead young monkey sewn on to the bottom half of a medium-sized fish. This was the mermaid that Barnum charged suckers to see, a terrifying twisted monkey-fish supposedly caught off the coast of Fiji, but more likely stitched together by a wily Fijian duping a British sailor.

The Feejee Mermaid

A note of warning. Gullible though the public may have been about the Feejee Mermaid, scientists can be too sceptical sometimes. Such was the fashion in those days for curious chimaeras from distant shores that when the first platypus was sent from Australia to scientists in London in 1799, they presumed this bizarre mammal with its fur, duck-bill and poisonous spurs was a fake. It turns out that evolution is even more ridiculous than fraudsters.

* Genesis 6:1–4 and Numbers 13:33 speak of the Nephilim, who over the years and in various translations have been interpreted as fallen angels, or the sons and daughters of heaven or of God, and often as giants. Given that the geographical scope of the Bible doesn't extend much beyond the Middle East, it's not clear what these unlikely giants were doing in upstate New York, but that didn't stop a number of people from discovering them in the seventeenth and eighteenth centuries. The Puritan Cotton Mather, who was also heavily involved in the Salem witch trials, believed that fossilized bones found in Albany, NY, in 1705 were those of Nephilim giants who had died in Noah's flood. They were in fact from a prehistoric mammoth.

Opening the file drawer

Everyone is vulnerable to confirmation bias. The consequences can be as trivial as horoscopes and psychics or as serious as undermining our democracies. Science, of course, should be above all of this. That is precisely the point of science – to over-rule these basic human errors, so we can see how things really are, rather than how we perceive them to be.

We have made a great fanfare in these pages of celebrating the fact that we humans, with all our psychological biases in place, invented science to give us a way of bypassing these very normal human errors. But research is done by humans and, hard as we try, shaking off the shackles of these inbuilt tendencies is not easy. And science, for all its august aspirations, is plagued by psychological biases, to its very core.

In just the past few years, we've begun to realize that the standards of science are perpetually troubled by confirmation bias and other human failings. The effects are so significant that we have a new name for a whole category of events and issues that arise when science falls prey to this human error. Collectively, these issues have become known as the File Drawer Problem.

Ignoring the boring

Along with confirmation bias in the suite of human weaknesses is a tendency to be overly fascinated by the new. We are finely tuned change detectors, which means that we love to overturn existing knowledge, and we elevate the novel over the familiar. These tendencies are termed *neophilia* – love of the new – and *theorrhea* – mania for a new theory. Sometimes, in designing experiments, we seek out the new and the intriguing, we pursue exciting new theories feverishly, and in doing so we ignore evidence that counters those shiny novelties.

The problem emerges when science falls foul of the same

BIAS: YOUR BRAIN IS COMMITTED TO TRICKING YOU

Confirmation bias is probably the best-understood cognitive bias, but more than a hundred others have been described, some trivial, some profound. Tick the circles for the biases that you think you have observed in you or someone you know (be honest!):

Present bias: An inability to make commitments to longer-term benefits that outweigh immediate ones. In experiments, people offered $150 today or $180 in a month's time tend to take the lower amount. Which isn't very clever. ○

The IKEA effect: People tend to put a higher value on furniture they have assembled themselves. ○

Belief persistence: Even after overwhelming evidence to the contrary, you continue to believe in a specific thing, such as an enchanted orchid or a magic time of day. ○

Opinion polarization: When presented with evidence that counters your opinion, you opt to believe your original opinion even more strongly. ○

The bandwagon effect: The tendency to believe something because other people already do. ○

Effective heuristic: Judging people's suitability for a task by superficial characteristics, such as tattoos or weight. Studies have shown that people tend to underestimate someone else's abilities for a job unrelated to body weight if they are overweight, and overestimate it if they are more typically proportioned. ○

Anchoring: In decision-making, the tendency to focus on one piece of information, usually the first one acquired, and ignore subsequent inputs. ○

The gambler's fallacy: 'I've flipped four heads in a row, so the chances are the next one will be a tail.' Nope. ○

Loss aversion: People prefer to avoid losing something than to gain the same thing. Which would have a greater impact on your mood: getting a £5 discount or avoiding a £5 surcharge? ○

Declinism: Everything in the past was better, and everything is basically getting worse. It wasn't. It isn't. Almost everything sucked much worse in the past than it does today, and most of you would be very dead already. ○

The Dunning–Kruger effect: Non-experts tend to confidently overestimate their abilities or knowledge of any particular subject. You'll know this one if you've spent any time on Twitter as a woman. ○

The rhyme-as-reason effect: Statements that rhyme are perceived to be more true. An apple a day keeps the doctor away, a false statement that must have been dreamt up by the Apple Marketing Board. It really doesn't. Or, in the murder trial of O. J. Simpson, where a glove was a crucial piece of evidence: 'If it doesn't fit, you must acquit.'

○

Bias blind spot: This one is a sort of meta-bias – it is the inability to spot your own biases, coupled with the ability to identify them perfectly well in others. We definitely don't have this at all. But you definitely do.

○

fallacies simultaneously. What happens to the experiments that are boring but produce reliable data? What happens to the ones that disprove a sexy result, or show that a big finding with a lot of impact is weaker than everyone thought? If those dull but worthy experiments failed to capture the scientific spotlight, perhaps they'd remain tucked away in researchers' filing cabinets, deemed not interesting enough to write up into a published paper, destined to be ignored and never seen again. How much of science is missing? This concern is precisely what prompted NASA scientist Jeffrey Scargle to start talking about the File Drawer Problem back in 2000.

We're piling on the issues here: there's the thrill of the new, ignoring the boring. And there's confirmation bias too: scientists are not immune. Are they seeking out data that support a conclusion rather than trying to qualify or test it?

One example comes from the world of ant research. In the ant world, nestmates are assumed to be less violent with each other than with ants from other colonies. That makes sense: an invader from another tribe is often unwelcome. Assessing aggression in ants is not easy, and involves making judgements about their behaviour – things like how they chomp their mandibles and rear up in a confrontation – and making judgements in turn brings in the potential for human error. A paper published in 2013 looked at 79 studies of nestmate aggression simply to see if the experimenters were biased in making their observations. In practice, that means: 'Did they know whether the ants they were looking at were nestmates or a mixture from different colonies?' The authors of the paper found that if researchers knew which ants were which, they were significantly more likely to report aggression than when the identity of the ants was unknown.

In medicine, double-blind testing is standard practice. This means that neither the experimenter nor the subject knows which drug is real and which is the completely inactive placebo. That way, neither the doctor nor the patient can influence the result. This should be an absolute standard in science, but, amazingly, many studies show that experiments were not blind when they could have been. It turns out that, in the entire history of studying ant aggression, less than a third of the experiments obscured the identity of the ants from the scientists.

Admittedly, assessing ant aggro might not seem like one of the most important experiments in history, but the fact is that there are many examples of exactly the same phenomenon in all branches of science, where failure to acknowledge our inbuilt biases, and to build in checks and

BLIND CHOPPERS

Blind testing, by the way, is not a new phenomenon. Possibly the first recorded blind test was instigated by King Louis XVI of France in challenging the claims of the French fraud Franz Mesmer – one of very few people whose surname became a verb, as in 'mesmerizing'. Mesmer believed in a special invisible form of magnetism, an energy field created by all living things that surrounds us, penetrates us, and binds the galaxy together,[*] that could be drawn out as a kind of vital fluid, sometimes stimulated by magnets. Louis recruited an esteemed scientific panel that included the great chemist Antoine Lavoisier and another rare surname-verbing Frenchman, Joseph Guillotin, inventor of the revolutionary chopper.[†] The panel tested the claims of mesmerists by asking them to identify vessels containing the mysterious vital fluids of mesmerism, while blindfolded. They failed.

[*] Yes, there's one of those film quotes.
[†] Ironically, in 1794, at the height of the French Revolution, Lavoisier met his death at the sharp business end of Guillotin's invention.

balances to counter them, massively skews the results.

The result of all of this is that science is biased. It's biased towards publishing dramatic results, novel results, results that affirm or enhance similar findings. It is biased towards big, positive statements.

Try this for a big statement: stand up, with your feet wide apart, chin out, and hands on your hips like Wonder Woman. Do you now feel confident, assertive, powerful?

Or a bit silly? Your reaction probably depends on whether you've seen the third most popular TED talk of all time.* In 2012, the psychologist Amy Cuddy gave a TED talk on 'power posing', based on a scientific paper published with her colleague Dana Carney a couple of years earlier. In the phenomenally popular talk – now viewed over 61 million times – Cuddy describes the effect that the Wonder Woman posture can have on your stress levels and confidence. It's a potent and compelling story. There is one small issue, though: it's not true.

No one has ever been able to reproduce Cuddy and Carney's findings, though plenty have tried. In 2015, Carney issued a statement saying that the work was flawed and the effect was not real; Cuddy, though, continues to promote similar notions, outside academia. The idea is still out there, and it's still popular, with politicians adopting power poses in attempts to appear more confident (though mostly they just look like they are wearing uncomfortable trousers, or maybe have mild rickets).

This is just one of hundreds of examples of similar high-profile scientific findings that are exciting, make big news and turn out to be not quite true. It's a phenomenon now called the 'replication crisis'. Psychology is particularly prone to this problem, but all branches of science suffer from it to a certain extent. Big findings, published in big journals, with big-name authors and lots of publicity, turn out to be weak, statistically fragile or sometimes just plain wrong, because science is designed, enacted, discussed and publicized by people. It's much less fun to repeat someone else's

---

* But it's not nearly as good as the sixth most popular TED talk of 2015, 'The Mathematics of Love'. You should definitely check that one out.

experiment than to do something bold and new, because it's time-consuming and expensive, and if they were right, they're still right and all you have done is confirm it. If they're wrong, loads of reputations and a lot of investment, both emotional and financial, are at stake.

But replicating results is an absolutely immovable cornerstone of the scientific process. Your results *have* to be independently verified by others. The founders of the Royal Society in the seventeenth century knew this. When they set up the first national scientific organization, a key element in what came to be known as the 'scientific revolution', the phrase they chose as its Latin motto was, and is, *Nullius in verba* – which roughly translates as 'take nobody's word for it'. We take that to mean that the true scientific method relies on sharing data and results so that they can be tested, repeatedly, and rejecting any assertion that something is true just because someone important says so. In practice, of course, we can't do every experiment from scratch ourselves – we have to rely on the presumption that the people on whose shoulders we are standing did an honest and good job, and that their work has passed muster. Contemporary scientific methods rely on checks and balances for this exact reason.

But it doesn't always work. The replication crisis is real, and is fuelled by cognitive biases. And it means that we sometimes generate and spread supposedly scientific truths that are simply not true.

We sometimes brag that the strength of science is that it is self-correcting. This is true, but only with the crucial caveat '*when we correct it*'. That correction doesn't happen magically, and science is not immune to human failures. That is because it is carried out by humans, and human minds are plagued with default settings that don't seek out the truth at all. We do have the capability to detect those biases and fix them, though – as long as we are aware of them. So, definitely be wary of people trying to con you. But be vigilant too, because your own brain is determined to trick you.

CHAPTER 8

DOES MY DOG LOVE ME?

Near this Spot
are deposited the Remains of one
who possessed Beauty without Vanity,
Strength without Insolence,
Courage without Ferosity,
and all the virtues of Man without his Vices.

With that beautiful epitaph, the early-nineteenth-century poet, celebrity, gambler and rampant sexual deviant Lord Byron paid tribute to his most beloved friend. This verse marked the passing of perhaps the only companion with whom he had an uncomplicated relationship, someone he cradled and nursed in his final days and hours. Byron had unconventional associations with men, women, family members and the tame bear he kept in his student digs at Cambridge (see box opposite). But truly, madly and deeply did he love his dog. The epitaph is in fact for Boatswain, his beloved Newfoundland, who died of rabies.

It's very clear that Byron loved Boatswain, and both of your guides in this book are also dog lovers. Jesse the whippet is the most recent addition to the Rutherford household, and Molly the cockapoo a long-standing member of the Fry family. We, and many others, love our canine companions unequivocally and profoundly.

What love is – whether lavished on a pet, another person or inanimate object – has been the subject of much scientific scrutiny for more than a century. What it feels like to love another has been pored over by philosophers, musicians, writers and dandy poets for far longer. But in this chapter we are setting ourselves a harder challenge: to answer the question not of whether we love our dogs, but whether they love us back.

Like all dog owners, the most frequent questions we ask our pets are: 'Do you want to go for a walk?', 'Are you hungry?' and 'Who's a good boy/girl?' To which the answers are 'Yes', 'Yes', and 'Is it me?' More sophisticated questions are somewhat harder. Do dogs feel joy, or sorrow, or doubt? Did Boatswain show no vanity, insolence, but only boundless courage? Are cats as contemptuous as their behaviour implies? Are foxes having good times when they screech in the night like babies being murdered? Whatever we might imagine is going on, we are fundamentally hampered when we dip into the science of animal emotions by one simple fact: they can't actually tell us how they feel.

Normally in science, we turn to animals because humans are a pain in the arse to study. Here we turn the tables, because humans are enormous bundles of emotions with the

BYRON'S SCIENTIFIC LEGACY

Lord Byron acquired his bear in protest at Trinity College rules that barred him from keeping his beloved dog there. Technically, Cambridge University did not specify bears in its statutes relating to lodgings and pets and so, being something of a smart-arse, he argued that they could not deny him a flesh-eating undergraduate ursine companion. This is a fairly typical saga for Byron, a man who was so debauched that even one of his lovers described him as 'mad, bad and dangerous to know'. His short life was packed with stories of misplaced bets, unusual sexual dalliances, adventures, rumours, obsessions and all the trappings of a rich life of melodrama and privilege.

And yet, while historians who idolize foppish men of questionable morals might disagree, we are very certain that Byron's greatest legacies are not his at all, and both relate to science.

First, his daughter, the brilliant Ada Lovelace, is widely credited with writing the world's first computer program, a mere century before a computer was actually built. She had spent many years working closely with the eccentric and exuberant inventor Charles Babbage (imagine Caractacus Potts in *Chitty Chitty Bang Bang* and you're pretty much there) while he was drawing up plans for his Analytical Engine. They had, in their hands, the designs for a fully functional Victorian computer that was the size of a cathedral and powered by steam. It would have worked, too, if only the duo had got funding from the government. Have a think about that: the world really was within a couple of signatures of a Victorian information age – with vast computers lining the Thames, connected by telegraph wires, gobbling up coal and spitting out punch cards.

Babbage and Lovelace both desperately wanted their computer to become a reality, but while Charles saw it as little more than a giant calculator, only Ada grasped the true significance of the invention. She knew that it had the power to go well beyond just a maths machine, and her ideas would go on to directly influence Alan Turing a hundred years later. In one particularly famous note she wrote: 'We may say that the Analytical Engine weaves algebraical patterns just as [a] loom weaves flowers and leaves.' Ada had realized that, one day, computers would help us to create art and music. She had predicted the invention of Spotify and Photoshop 150 years early.

Byron's second great legacy began its life in his holiday home on Lake Geneva in 1816, when he challenged the 18-year-old Mary Shelley to come up with a ghost story. According to her own journal, after a few tumultuous nights in which the Moon poured through her shutters (this is verified by astronomical archaeology in 2011; the Moon was indeed unusually bright in Geneva on 16 June), Shelley concocted a story about the creation of a monster from the body parts of executed criminals, invigorated by the newly discovered galvanism, whereby electricity would twitch the muscles of the recently deceased. That story was published in 1818 as *Frankenstein*, arguably the first – and inarguably one of the most significant – works of science fiction. Mary Shelley, for what it's worth, was not averse to a bit of Byronic melodrama: she lost her virginity to her future husband Percy Bysshe Shelley in a graveyard, and kept the ashes of his heart in a silk scarf wrapped in a page of his poetry. That, in our considered opinion, is well goth.

key advantage of language, and our inner states have been the subject of literature and science for the last few thousand years. We can tell each other when we are sad, angry, happy or bored. We can use art or words (or, heaven forfend, mime) to express what is inside our minds and souls. Our culture depends on expressing ourselves in such a way that others can understand how we feel, and in the last couple of centuries science has jumped on the cultural bandwagon to try to understand how our feelings manifest themselves in our bodies. But even here, the definition of emotions – those internal biological states within the nervous system that relate to our feelings – are strangely non-specific. You might have thought that we would have pinned this down by now, but it is a subject still fraught with disagreement and argument. There is currently no scientific consensus on a definition of what an emotion is, in humans or otherwise. So, while we hope you're feeling brave, apprehensive and excited, but not anxious, afraid or frustrated, we can't guarantee that by the end of this chapter you won't be.

The experiment when Darwin's French friend electrocuted an old man's face

1872, and Charles Darwin was on a roll. In the previous 13 years, he'd published two of the most important works in the history of humankind, and now it was time for him to drop his third blockbuster book: *The Expression of the Emotions in Man and Animals*. It was a bestseller. Audiences were enthralled by Darwin's attempts to understand the internal emotional states of animals (including us) and how they might relate to the outward expressions of their faces and bodies. There is an entire chapter on blushing. Another on weeping. He writes about cows who 'throw up their tails in a ridiculous fashion' when they 'frisk about from pleasure'. There's a sweaty hippopotamus who wasn't having the best time in labour, and an impatient horse. And there's page after page of Darwin looking very, very carefully at the faces of monkeys.

In one notable passage, Darwin describes a trip to London Zoo, where he introduces a freshwater turtle to a monkey and watches as the monkey's eyebrows rise in astonishment. He dresses up a little doll to show to a crested macaque, and writes that the ape stared intently at the toy with wide-open eyes, which he interprets as terror. These, Darwin argues, are human-like emotions as experienced by primates.

Much earlier in his career, back in 1838, Darwin had spent some time with an orang-utan called Jenny and noticed that her behaviour was very much that of a small child. When a zookeeper withheld an apple from her, Jenny 'threw herself on her back, kicked & cried, precisely like a naughty child.— She then looked very sulky & after two or three fits of pashion [sic], the keeper said, "Jenny if you will stop bawling

& be a good girl, I will give you the apple."— She certainly understood every word of this, &, though like a child, she had great work to stop whining, she at last succeeded, & then got the apple.'

The expressions and reactions of apes often seem to mirror our own, but Darwin does note that he's never seen an orang-utan frown. Neither have we, come to think of it.

Despite all the talk of monkey tantrums, Darwin's book is primarily concerned with what all this means for humans. By studying animal expressions, he is testing out an idea: that evolution has stumbled upon emotions as a way for creatures to communicate their internal states to others, that by outwardly expressing how we feel, our faces and bodies provide access to our innermost feelings, to our very souls.

Darwin then carefully considered what all this meant for the principles of human emotions, and went to some lengths to determine what kinds of expressions a human face is capable of making. He examined photos of actors posing with characteristic gurning faces to express anger or disgust, and pictures of babies and children pulling faces of happiness or a sneer.

He also tried to rationalize the musculature to make those faces, and for that he enlisted the help of the French scientist Guillaume-Benjamin-Amand Duchenne de Boulogne, best known for giving his name to the terrible disease Duchenne muscular dystrophy. Like Darwin, Duchenne was fascinated by the contortions of the many facial muscles within human expressions, and had resolved to try to test them out in the most scientifically vigorous and inventive way he could think of. Not by telling a few jokes to provoke a smile. Not even by showing them a freshwater turtle either. No, Duchenne decided to examine idealized facial expressions by electrocuting an old man's face.

The subject of the experiment was, in Duchenne's own words, an 'old toothless man, with a thin face, whose features, without being absolutely ugly, approached ordinary triviality'. It's generally not the done thing to write that about your subject participants, but that's the 1860s for you. They did get the man's consent for the experiment, at least. And so Duchenne held two electrified metal probes to different points on his face to isolate the muscles involved in various expressions. There are 42 such muscles in total, which combine to give every possible glance, wink, grimace, smile and smirk on the human face. Duchenne examined the furrows on the skin as the muscles were contracted, building a catalogue of genuine facial contortions. The expressions, fleeting as they were, were captured using photography, which had only been invented a few years earlier. The nameless, toothless man does not look like he's having a particularly good time. Duchenne, however, looks like he's having a ball.

These photos, along with those of actors and children, were probably the first to be printed in any book and formed much of the basis for Darwin's thesis on human emotions.

Looking through dozens of them led Darwin to conclude that some emotions are common among many living creatures, and that we each express the same state of mind by the same movements. There's the strongly marked frown of anger. The purposeless dance of joy. The open-eyed look of terror. Darwin boiled down the complexity of human feelings to six universal emotions that could be felt by all: anger, fear, surprise, disgust, happiness and sadness.

During the 2004 Paralympic Games, the American psychologists David Matsumoto and Bob Willingham found a new way to study the science of emotion, by carefully watching what was happening on the judo mat. During the gold- or bronze-medal matches, a high-speed camera captured the expressions on the faces of the athletes. Joy spread across the flushed cheeks of the winners, and sadness or contempt on the faces of those who had lost, but what Matsumoto was looking for was a difference in expression between two key groups of athletes: those who could see, and those who were congenitally blind.

In their moment of triumph, judo champions showed the same facial expression regardless of whether they were sighted, had become blind in life, or had been born sightless and never seen another human face. Some of the athletes would never have observed happiness on the face of another, and yet, across the board, the photos showed how the winning athletes' zygomaticus major (the muscle connecting the corner of the mouth to the cheekbone) contracted, giving them a full, broad smile from ear to ear. And their orbicularis oculi (the muscle surrounding the eye socket) engaged, pulling up their cheeks and causing a slight half-closing of their eyes into the most genuine gleeful expression of joy.

Here is a sporting demonstration of Darwin's great idea: that emotions are an ancient and innate part of human nature. If the judoka were anything to go by, our emotions are universal and it is our faces that hold the key.

This too is the theory of Paul Ekman, one of the silverbacks of twentieth-century psychology. Since the 1960s, Ekman has spent his entire career attempting to formalize Darwin's ideas and trying to sciencify the concept of core universal human emotions. In what was perhaps his most influential experiment, he took pictures of actors' faces contorted into frowns or grins, or wide-eyed in surprise, or wincing with pain, and showed them to people around the world. These included remote tribespeople from Papua New Guinea who had had little contact with people from Western cultures. What Ekman reported was that a smile was a show of happiness to everyone, and the look of terror could be spotted by anyone, no matter where they were from. The Papua New Guineans could reliably identify the emotion behind the expressions. Ekman concluded that our facial muscles contort to reflect fundamental emotional states, the same as those outlined by the grand master of all biology himself, Charles Darwin.

This is a theory of emotion that feels right. It seems intuitively correct. Children are taught to recognize the smiles and frowns of happy, sad and angry faces. In the twenty-first century we gave each of the emotions their own emoji, and once social media sites built buttons letting us react with an emotional sticker, they have become a shorthand for

communicating. The Pixar film *Inside Out* is based on the idea
that the universal emotions (apart from surprise, surprisingly) are
distinct characters within our minds. These Darwinian ideas are
locked into our culture: it is universally acknowledged that our
emotions can be categorized and that the categories can be read
from our expressions.

There's just one problem with this theory. It isn't true.

🫤 Can't read my p-p-p-poker face

There are a few issues with this classical view of emotions, but
let's start with the most obvious. Your face doesn't always tell
the truth. You don't have to be a monk or stony-faced poker
champion to know that controlling your facial expressions
is a way of hiding how you feel inside. The reason we think
this should be obvious is because there is a multibillion-dollar
industry that dominates culture, filling our lives with joy, sadness,
thrills, laughter and terror on a daily basis, and it is entirely based
on this disconnect between what we feel and what we display.
It's called show business. Actors are paid to pretend, to convey
emotions that they are not actually feeling. When Jack Torrance
axes a hole in the bathroom door in *The Shining*, intent on
murdering his terrified, screaming wife Wendy, Jack Nicholson
wasn't filled with ire from the spirits of long dead Native
Americans, Shelley Duvall wasn't really fearing for her life, and
after the scene was shot they both laughed. Luke Skywalker's
contorted, weeping face and impassioned 'NOOOOOO!' at
finding out that Darth Vader is actually his dad is no less powerful
with the knowledge that Vader was actually the body builder
David Prowse speaking with a thick and quite high-pitched West

Country accent.* And as for *When Harry Met Sally*, we're fairly sure that Meg Ryan wasn't having a loud orgasm in a café when her character Sally was faking it to prove the point that we can't truly know anyone else's inner state, no matter how confident a man is that he has satisfied a woman.

Irony upon irony, this whole house of emotional cards is built on the fact that both Darwin and Ekman *used actors* to test the existence of basic human emotional states. The pictures that the Papua New Guineans saw, like many of those in Darwin's emotions book, showed people who weren't reflecting their internal emotional state at all but were pretending to, coached by the scientists who had decided what the norms of those faces should look like.

And yet, when it comes to those norms, do you really buy the idea that happiness or sadness or disgust must look a certain way? Think about your favourite films and your favourite actors. The very reason that Lupita Nyong'o, or Meryl Streep, or Al Pacino, or Denzel Washington, or Helena Bonham Carter or any of the great actors are as astonishingly good at conveying the complexities of an emotional moment is precisely because they *don't* pull faces that we are told are instantly recognizable as core basic emotions. When was the last time someone won an Oscar for scowling when angry? Or pouting when sad? Roger Moore, lovable though he was, is primarily known for being a bad actor because his emotional range included raising an eyebrow and not

* David Prowse was the man in the suit, and also delivering the lines on set. James Earl Jones's sonorous Vader voice was dubbed in the edit. On set, Prowse actually says :'Obi Wan is your father', because they wanted to keep it a secret even from the cast and crew. We realize we have spoiled this amazing scene in this footnote, but if you haven't seen *The Empire Strikes Back* yet, what the hell have you been doing for the last 40 years?

much else. When Rick says to Ilsa that she should forego their deep love and get on the plane for the sake of the fight against the Nazis in *Casablanca*, Ingrid Bergman doesn't make an 'ooh' shape with her mouth and widen her eyes like an emoji. She doesn't do that because only a terrible actor would do that. Or a clown. And *Casablanca* with clowns is not a film that should exist.

😳 Surprise!

And yet we still cling to these faces, an emoji representation of how we feel. Surprise: eyebrows raised, wide eyes, an 'ooh'-shaped mouth or, for a real shocker, a dropped jaw. Everyone knows what face we pull when surprised. Why would science even need to test this?

Well, because it's not quite true. We are hoping that this comes as a bit of a surprise to you right now, because it's such a ubiquitous notion. Are you doing the face? We suspect not, because it turns out that in surprising situations the vast majority of people don't make the stereotypical surprise face at all.

A shocking experiment took place in Germany in 2011, when the psychologists Achim Schützwohl and Rainer Reisenzein decided to really freak people out and observe their faces during this outfreakery. Unwitting subjects were brought into a dimly lit room, given soundproof headphones and told to listen to a four-minute audiobook by Franz Kafka. The story, 'Before the Law', concerns a man trying to persuade a doorkeeper to allow him to pass, only to be denied, for years, until the man eventually dies. It's a parable. We don't understand it.

The subjects listened to the story, as decreed by the instructors, who waited outside until they had finished, whereon the subjects left the room, thinking they were going to be asked some questions about the story (it having been presented as a memory test). During the four minutes, however, Schützwohl and Reisenzein were outside the door speedily constructing a new, brightly lit green room, unfurnished but for a single red chair with a stranger sitting on it. As the unsuspecting subjects opened the door, instead of being in the corridor they came in by, they were in a completely different space, with someone weird staring them down for 15 seconds.

In all cases, the subjects reported afterwards that this WTF moment was indeed extremely surprising. They also thought that they had expressed their surprise on their faces. But they were being videoed, and their actual facial expressions coded according to Ekman's theory. It is pretty weird to walk into a room via a corridor and, on walking out, be faced with a stranger in a strange room, but only one-fifth of the participants pulled anything like the classic surprise face. The room was specifically designed to elicit what was presumed to be the evolutionary reason for making that face in a novel situation: the sudden and unexpected switch from dim to bright light would make the eyes widen in order to let the new information flood into the brain, just in case it was threatening. But no. Even when they had a surprise encounter with a friend in the corridor instead of a stranger, the number only rose to a quarter. The vast majority of people just didn't pull that face.

Similar results were found in the study undertaken by Matsumoto and Willingham on the Olympic judo mats. It's true that there was no statistical difference between the spontaneous expressions of sighted and congenitally blind competitors, but that doesn't mean everyone made the same face. The muscles may well have contracted to raise the cheeks

THE INDUSTRIALIZATION OF FACIAL EXPRESSION

Despite the increasing number of researchers who are calling into question Ekman's ideas of discrete, universal human facial expressions, emotion recognition has become a hot topic in the world of machine learning.

With machine vision now able to capture and classify facial expressions according to Ekman's theory, Disney films are tested for the emotional reactions they provoke before release. By screening them in cinemas with cameras pointed at each audience member's face, an algorithm analyses the images as they capture the audience's reaction, to see if people are laughing and crying and paying attention in all the right places. One comedy club in Barcelona, using a similar idea, even decided to forego the entrance fee, instead charging each audience member 30 cents per laugh, capped at €24 across the course of the evening.

There's little at risk when the stakes are only a dodgy Disney sequel (looking at you, *Frozen 2*) or an expensive comedy ticket, but the use of Ekman's emotional classifications has a pernicious side when teamed with the automated authority of artificial intelligence.

Within airports, cameras trained on passengers are now on the lookout for anyone displaying stereotypical signs of guilt and suspicion. There is a Hong Kong-based start-up which sells technology to schools; they claim that it can keep an eye on whether or not students are paying attention in the classroom. In some hospitals around the world, these algorithms are beginning to be deployed to determine the 'true' pain levels of patients with chronic conditions and decide if they need medication.

And all the while, the algorithms these technologies are using are based on images made from actors' impressions of an emotion – actors who are coached into contracting and relaxing a combination of facial muscles according to scientists' agreed definition of what that emotion should look like. But we know that, by and large, these stereotyped expressions are not found on real faces. One meta-review of over 1,000 studies published in 2019 found that, on average, we make the Ekman faces only 20–30 per cent of the time. No matter how fancy technology is, science cannot help you to reliably infer how someone feels from how they look.

and half close the eyes in the smiles of contestants regardless of their vision, but it occurred in only 37 of the 67 sighted players and 7 of the 17 who had been blind from birth. Less than half of the sighted group let their jaws drop (it was a little over half in the blind group), and the muscle that upturns the mouth into a smile only did so in 30 of the 67 sighted athletes, and 11 of the 17 who were born blind.

It's not that people don't smile when they're happy. They do, sometimes. Maybe often. But try to declare a rule that says smiling equals happiness (or that happiness equals smiling) and you'll come up short. The stereotyped expressions championed by Ekman and co. are just that. These are not rules that work for everyone. Expecting the surprise face or the disgusted face or the happy face to appear every time someone is surprised or disgusted or happy is a bit like trying to define birds as anything with wings that can fly. It's a description that feels like it should do the job, until you think about it a bit harder and remember dodos and bats and ostriches and bees, and realize that biology is a whole lot more complicated than you thought.

So what does this mean for the theory of universal emotions? What of Ekman's celebrated findings with the tribespeople of Papua New Guinea, who could so expertly identify the emotions in photographs of actors' faces?

In the twenty-first century, the researchers Carlos Crivelli and Sergio Jarillo embedded themselves with the Trobriander tribe, a population of fisherpeople and horticulturalists who are relatively isolated even from other Papua New Guineans. Crivelli and Jarillo learned their language, Kilivila, and took Trobriander names: Crivelli became 'Kelakasi' and Jarillo 'Tonogwa'. When they used Ekman's photos in 2016 and repeated the experiments with the Trobriander, they got completely different results. Overall, they agreed with Ekman's predictions less than a quarter of the time, meaning that the Trobriander didn't recognize facial expressions that we associate with the so-called basic emotions. Happiness tallied with Ekman's earlier results quite well (but not perfectly). Others were much more variable. The face that we in the West most associate with fear – wide-eyed, lips-parted and gasping – the Trobriander mostly interpreted as 'angry'.

The fact that Crivelli and Jarillo spoke Kilivila was crucial in challenging the earlier results. In his original studies Ekman had used a translator, and presented a list of possible emotions with the photos which may well have primed his test subjects – that is, inadvertently steered them towards an answer that they might not otherwise have given. When looking in confusion at a photo of a man's face, the translator nudged them along by giving a little backstory to the photo – this man's child has just died; what expression is he making?

The Trobriander are not alone in finding it easier to judge an emotion when they're given the story around it. Hollywood has known for some time that it's the context around an image of someone's face that makes all the difference (see box below). The same is true with Facebook emoticons. The neuroscientist Lisa Feldman Barrett found that people could identify them more easily when they were given a set of emotion words to choose from. Without the prompts, people could spot 'happy' and 'surprised' but were no better than guessing at the others.

Back in Papua New Guinea, the most common response to all the photos wasn't a definitive core emotion – it was 'I don't know'. After years of being the subjects of some pretty dodgy science, maybe the Trobriander just wanted all the researchers to leave them alone.

HITCHCOCK, THE MASTER OF SUSPENSE

Alfred Hitchcock is widely and correctly regarded as one of the greatest film directors of all time, and a master of manipulating our emotions; he was quite explicit about his intention utterly and shamelessly to do so with the power of suspense, music, drama and horror. *Psycho*, *North by Northwest* and *Vertigo* are among the best films made in any genre, not least because of his ability to toy with our emotions, especially when it came to suspense: 'There is no terror in the bang, only in the anticipation of it.'

Cinema at its best is an emotion-manipulating machine, and Hitch understood its power. In a short film in 1964 (featuring a sharp-suited cigarette-smoking interviewer – it was the 1960s), he explained how interpretations of facial expressions were dependent on the context.

He showed a clip of himself, fat, jowly and stern-faced, which then cuts to a mum playing on the grass with her toddler, and then back to Hitch's face, which cracks into a smile. 'Now what is he as a character? He's a kindly man.' He then showed the exact same shots of his own face, but instead of the mum and child, the middle clip is of a young woman in a bikini. 'What is he now?' says Hitch. 'A dirty old man.'

This short, simple edit demonstrates clearly how context-dependent our facial expressions can be. Hitch flips from being kindly to a pervert based only on what we perceive he is looking at. It shows how easy it is to manipulate a viewer by priming them, which is what can happen, and did, in Ekman's experiment.

Hitchcock, for the record, *was* a dirty old man. Sadly, his ability to manipulate our emotions on screen was matched by his infamous obsession with, and horrible abuse of, his leading female actors, some of whose careers he ruined out of little more than spite.

Emotional range

Our inner states are not in step with our outer expressions. Our faces are mobile and demonstrative. We contort them to express ourselves, pulling and tweaking those 42 muscles (humans have more than any other creature with a face) into the mugs that we can all recognize and interpret. The subtlest shift in one tiny muscle can mean the difference between come-to-bed eyes and go-to-bed exhaustion. And while people often don't necessarily show their emotions on their faces, the reverse is also true: deciphering an outward expression does not mean you can determine what's going on within.

But there's a second, deeper issue with the Darwinian view of human emotion, one that challenges the idea that our complex maelstrom of feelings can be distilled down into six simple categories. This goes beyond the simplistic – and debunked – notion that our faces are windows to our feelings, and concerns itself with the simplicity (or otherwise) of our internal states. It's an issue that becomes evident in the words we use to describe how we feel.

Language betrays the nuance of our emotions – it lays bare the subtleties, the imperceptive differences, the refined ability we have to describe how we feel. The English have a reputation for repressing their emotions, which of course doesn't mean that they don't feel them, stiff upper lip and all that. But there are plenty of emotional states for which the English simply don't have words. Well-known examples are *Schadenfreude*, German for the pleasure experienced in someone else's misfortune, and *l'esprit de l'escalier*, French for the frustration of thinking of a zinger, a perfect comeback, seconds too late. Here are some of our favourites:

Iktsuarpok (Inuit) – displaying the excitement of waiting for someone to arrive by repeatedly looking out of the window.

Natsukashii (Japanese) – the bittersweet sense of happiness for something long gone, and sadness that it will not return.

Saudade (Portuguese) – a deep emotional state of nostalgic or profound melancholic longing for an absent something or someone that one loves.

Desbundar (Portuguese) – to shed inhibitions when lost in joy. Basically, to dance as if no one is watching, but not to be confused with . . .

Mbuki-mvuki (Bantu) – to shake off your clothes and dance uninhibitedly. Which might lead to . . .

Pena ajena (Mexican Spanish) – the cringe of embarrassment when witnessing someone else's humiliation.

Gigil (Filipino) – the irrepressible desire to cuddle or squeeze something that is unbearably cute.

Feierabend (German) – the party-time mood at the end of a working day.

Ei viitsi (Estonian) – I really can't be bothered to do anything, or even get off the sofa.

Yugen (Japanese) – the sense of awe and wonder at discovering the mysteries of the cosmos, often best expressed simply with the word 'dude'.

All these gorgeous words, and many more, demonstrate that human language evolves to describe how we feel. They are learned, culturally specific and context-dependent, and yet we are in no doubt that you, readers, recognize each of those emotions even without having known the word for them. Humans are wonderfully, bewilderingly, frustratingly complex, and, despite science's best efforts, emotional states refuse to fit neatly into well-defined boxes.

This is the best explanation we can offer for why the Darwinian view of discrete emotions is inadequate. The reduction of emotions to core ones is not how we feel. Attempts to define basic emotional states do not reflect the fact that happiness, anger, sadness, fear and surprise are not irreducible, are not universal and are anything but simple. These are not Lego bricks of feelings, nor are they the subatomic particles of emotions. Happiness is a broad collection of positive emotional states. You might be happy because someone told a funny joke, or you just won a bet, or one of your friends or family achieved a life goal, or scored an actual goal, or you were cleared of cancer after chemotherapy. Those happinesses are not all the same, and do not carry the same weight or personal significance, but we broadly label them all as happy because if someone asks you 'How are you feeling?', a simple, socially acceptable and concise catch-all answer is 'I feel happy'. You may never be happy, or scared, in quite the same way twice. Is the fear of missing a penalty the same as that of being diagnosed with cancer? Of course not, but you may well describe both situations by saying, 'I am scared.' We don't normally contextualize the complexities of our internal emotional states, and we learn to use language shortcuts.

The science that has prevailed in trying to understand

emotions seems to have forgotten that. Darwin started the ball rolling by suggesting that there were just six basic emotions; and others, notably Ekman, picked up this ball and ran and ran with it. During the process – which dominated much of the research into emotions in the twentieth century – we all missed the starting point, which is that emotions are complex, and that our language simplifies their complexity out of necessity.

We are scientifically and emotionally all in a pickle. You know what that phrase means because you have learned to use it as a simple short cut for an emotional state that includes confusion and complexity and frustration. And pickles are nice.

Other animals

Not for the first time in this book, studying humans has proved more than a little frustrating. Despite our desire to understand and express our inner states with language, the science of human emotions is still a bit of a mess. The original question, however, wasn't about us – it was about our dogs and whether they feel love.

There is little doubt that animals do feel basic emotional responses. Although we might not be able to determine precisely the innermost feelings of another being by their outward expressions, we can nonetheless be confident that many animals do express what we term fear, for example. Many will widen their eyes when faced with danger, perhaps to maximize their vision to put them on a war footing. You'd have to have a heart of icy granite to think that a purring cat or a dog wagging its tail is anything other than clear and unequivocal evidence of joy, and both will scrunch their eyes in what we reasonably assume is pleasure. In Darwin's emotions book, he notes that baby orang-utans giggle when tickled, and chimps bark when happy – he couldn't distinguish between joy and affection in their faces, but did observe that their eyes 'sparkle and grow brighter'. He notes that Duchenne (he of the old man's face experiment) kept a very tame monkey in his house, and that the monkey would raise the corners of its mouth when offered a delicacy, like the muscles of a smile do.

And what of more complex human-like emotions as experienced by other creatures? Do animals get angry? They certainly express the threat of violence, which looks similar to anger, but here we come up against the problem that the way in which we describe emotions is very human-focused. A dog, wolf or chimp will bare its teeth when issuing a threat, and may even engage in a frenzied attack, but whether that counts as a fit of rage or a controlled and considered reaction to danger is impossible to know.

There is some anecdotal evidence that elephants and some of the great apes grieve following the death of a close family member. Both are known to remain close to the body, in a way that resembles both sadness and mourning. Most notably, there was the heartbreaking story of Gana, an 11-year-old gorilla in Munster Zoo in Germany, who in 2008 became famous after newspapers printed pictures of her cradling the lifeless body of her infant child.

But finding reference points for understanding the outward expression of feelings in animals can be hard. No other animal has such complex facial musculature as humans. Dolphins, notably the bottlenose, have permanently upturned mouths and wide eyes; this makes them look smiley and jolly all the time, despite the fact that they are the brutal, viciously violent king bastards of the oceans.* A dolphin has no facial muscles with which to change its expression, and therefore cannot convey any emotion other than a stupidly chirpy face.

Regrets, I've had a few

There are, however – just occasionally – scientifically valid ways to test whether animals are capable of more complex emotions. Does a squirrel feel embarrassment if it drops a nut in front of other judgey squirrels? That one we don't know. Do crocodiles feel guilt after they've dragged an innocent yet tender baby deer to the bottom of a lake? Also, don't know. Does a rat feel regret at choosing a less tasty food, only to discover that if it had waited a bit longer it could've chowed down on its favourite? The answer is: absolutely yes.

There are a couple of things you need to know about lab rats. The first is that, even though rats have a reputation for eating any old rubbish, including things that we would not class as food, they also have certain taste preferences. Standard rat food in experiments includes flavoured pellets infused with banana, cherry, chocolate or other flavours; some rats like banana, some like cherry. They're not *that* picky and will happily eat either or both, but, given a choice, they definitely have a preference. The second thing worth noting is that rats are intelligent, learn quickly and are very trainable. The promise of food is a major incentive for getting a rat to do things that will

* Upsetting though it might be to hear, bottlenose dolphins will bite, bash and kill infants; teenage male dolphins form gangs to kidnap young females and prevent them from escaping by biting, tail-swiping and killing them. Flipper has a lot to answer for.

help your experiment: you can train a rat to press a lever to get a food reward or negotiate a maze.

The psychologists Adam Steiner and David Redish from the University of Minnesota have pioneered an experiment to test a very specific and complex emotion in rats, and they did it by building them a small food court. It's called Restaurant Row, and is basically an octagonal arena with four food options in opposite corners. Imagine a dining area in an airport or shopping mall with a burger joint, a pizza parlour, a sushi bar and a fish and chip shop. You'll eat any of them in a hurry, but your absolute favourite is a burger.

Today there's a queue for the burgers, but the sushi and pizza are already prepared and require no wait. So you cut your losses, head for the pizza and grab a slice of pepperoni. But just as you tuck into your slice, you notice that the burger queue has gone and it's too late. You're two bites in and fully committed. You promise yourself you'll be more patient next time.

What are you feeling now? The answer is probably regret. It's an explicitly negative emotion that is more than simple disappointment. Regret is self-reflective disappointment, coupled with an implicit promise to do better next time – 'If only I had been patient, I could've had a juicy burger.'

That is precisely what Restaurant Row tests, but in rats. They used rats with a preference for one flavour over the others, say cherry. Then they trained them to associate a bleep that falls in pitch with waiting time for the release of some food. As the rats navigated Restaurant Row, they could hear the pitch of each offering and decide if they wanted to wait or move on, but if they left, the offer of that particular flavour would be withdrawn. Rats don't like waiting, but they were prepared to hang around longer for some pellets (like the tasty cherry morsels) than others. The scientists set up the experiments so that sometimes a rat, faced with a long wait at its favourite establishment, would move on to its second or third choice hoping for a shorter wait, only to realize that there was an unsatisfactory wait there too.

Mean-spirited though it sounds, it did make some rats mildly unhappy, which is perfect for studying animal emotions. At this point, you may be wondering how you can tell that rats experience regret, and – quite reasonably – how the scientists could tell the difference between a disappointed rat and a regretful one. Rats don't have a great facial range, so they don't look wistful and forlorn, and post-experiment interviews failed to yield any useful information. But they do something different – and telling – in the two situations. When regretful, they spent some time turning their heads and looking over to the food they could have had if only they'd waited. When merely disappointed by the reasonable decision to move on because the wait was longer than they were used to, they didn't turn back. And, more importantly, they learned from their impatience. The next time round, the rats waited it out to get their preferred cherry; the gamble hadn't paid off, so they played more cautiously.

This is the function of regret – to learn from our mistakes. It's a very sophisticated emotion to express. You have to make calculations of the odds for getting the best possible outcome, and recognize that your calculations went awry. You have to process what has already happened, and predict what will happen next time if you react differently.

We are in the era of brain-scanning, too, and this is not limited to humans. Steiner and Redish peeked inside the rats' skulls while they were perusing Restaurant Row and recorded activity in an area of the brain that we already know is active when humans express regret. There, in the rats' orbitofrontal cortex, Steiner and Redish saw specific cells flicker with life according to different flavours and certain restaurants, so they had a picture of neural activity in each specific scenario. When the regret scenario played out, they saw cells spark that represented the restaurant the rats had passed up. Their brains were alight with memories of the moment when they had made the wrong decision. The rats who liked cherry were still thinking about cherry when they moved on and got banana.

We are acutely aware that we are now being reductive in just the way that we have already admonished. Regret is a hugely complex emotion, and we can't really know for sure what the rats were feeling, because none of them have broken out into a Frank Sinatra tune when lamenting their lost cherry. But this bizarre, very clever and slightly mean rat-teasing experiment does show that at least one animal has at least one complex emotion that is comparable to a human feeling.

Is this love?

Of all the emotions, the greatest is love. At least that's what Paul's letter to the Corinthians says, as repeated at 96.4 per cent of weddings.* Pretty much every writer, composer and musician has had a stab at defining love over the last few thousand years. The 1980s poodle-haired metallers Whitesnake asked, 'Is this love?', and in the 1990s one-hit wonder Haddaway broadened the question out to 'What is love?', but neither even attempted a comprehensive answer. Dolly Parton and Whitney Houston declared that they will both always love an unspecified person, but failed to detail the mechanism by which they plan to defy time's absolute tyranny. Anne Hathaway's character in the epically confusing science-fiction film *Interstellar* picked up on that theme, declaring that 'Love is the one thing we're capable of perceiving that transcends dimensions of time and space. Maybe we should trust that, even if we can't understand it.' However, she didn't show her working, so we have to park this in the file marked 'In need of peer review'.

Let's see what Wikipedia has to say on the matter:

Love encompasses a range of strong and positive emotional and mental states, from the most sublime virtue or good habit, the deepest interpersonal affection, to the simplest pleasure.

And they say romance is dead. As with much of our language of emotions, the meaning of love is not only imprecise but context-dependent. Both of this book's authors deeply love pizzas and have consumed many during its writing, but we are confident that it is not the same emotion that we have for our friends, or parents. Nor is it the same as the love of our

---

* We made this stat up, but it sure feels like it. The full quote, as if we need to repeat it, starts with 'Love is patient, love is kind. It does not envy, it does not boast, it is not proud. It does not dishonour others, it is not self-seeking, it is not easily angered' and goes on like that, which is rather lovely, of course, and useful in trying to define this nebulous but powerful emotion. However, the earlier King James version doesn't mention love once. Instead, it talks of charity: 'the greatest of these is charity'. You don't hear that at weddings. It also contains the oft-repeated phrase 'When I was a child, I spake as a child, I understood as a child, I thought as a child: but when I became a man, I put away childish things.' As is probably very apparent to you if you have made it this far into this book, neither of us have successfully managed that.

children, whose breath is like a beating heart, nor is it the love of our partners, which is frankly none of your business.

Experiments with fMRI brain-scanning show that when we are given pictures of people we are romantically attached to, certain areas of the brain become notably active, including structures such as the amygdala, the hippocampus and the prefrontal cortex, all of which are associated with pleasure. But they're also active during sex, eating and drug use. Indeed, it's difficult to scientifically distinguish between how we feel about various stimuli: the dopamine hit when eating an amazing slice of pepperoni pizza is a key component in feeling good, but it's shared with the very same neurochemical pathway when we fall in love, have sex or even when exercising, and maybe all three at the same time. Or four if you're really into pizza.

Dopamine is one of the chemicals associated with pleasure that flood our brains. It's the one that helps explain the throbbing heart, sweaty palms, anxiety and passion that come in those early days of love. But the same could be said for getting drunk. Indeed, a study of fruit flies in 2012 showed that males which had been sexually rejected by females went on to drink four times more alcohol than those who got lucky, and were stimulating the same neurochemical pathways.

Despite the ever-finer techniques we have for peering inside our bodies to understand what is happening when we feel love, or indeed any emotion, there is no signature, no fingerprint, no barcode that says 'This person is loved up'. The same applies to dogs. Emotions are hard. There isn't a light that pings on in their brains which indicates that they can feel love. But that is the challenge we have set ourselves. We love our dogs, but do our dogs love us?

Do our dogs love us?

Dogs have been part of our lives for tens of thousands of years. The science is not settled yet, but it's thought that dogs evolved from a form of wolf that is probably extinct, rather than from the lineage of present-day wolves. Our best theory for how this happened is that the less timid members of these packs began hanging around human settlements, receiving scraps and leftovers when food resources were hard to come by during the peak of the Ice Age.

We don't know exactly what these wolves looked like, but they were probably smaller than the wolves of today. We have contested evidence of ancient dogs from at least 36,000 years ago, and definitive bones from a burial site in Germany that is around 14,000 years old. The remains include a jawbone: the teeth are less crowded than a wolf's, the snout is shorter, which we think reflects selective breeding by humans to reduce aggression in them. We think that these dogs were used for hunting, as well as being companions; small, fast hounds could bound through forests too dense for us to navigate.

We domesticated dogs before we were farming, and they've been our best friends ever since. Modern versions of dogs are of course hugely varied and lovely; we've bred them over the last few centuries, and their changing facial anatomy is very obvious – from the snub nose of a pug to the achingly handsome, long face of a whippet, of whom the best-looking of all is Jesse Rutherford. Those faces, adorable as they are, have been bred by us, for looks, for function, or a bit of both. But we now know that we have inadvertently selected facial characteristics for the dogs to communicate with humans.

All dogs, from Cavachons to Rottweilers, have a muscle

that allows them to move their eyebrows that is completely absent in wolves. The eyebrow movements of dogs generate what is known as paedomorphism, that is, looking like a young pup or even a baby, and this encourages us to empathize with and nurture them, as we would a baby. The sad look a dog can give you is part of a genetic repertoire that encourages you to love it. Puppy-dog eyes are our own creation.

We have also changed dogs' brains through breeding according to what we want from them. In 2019, scientists decided to apply some of the techniques we've been using on ourselves for years to have a look inside the heads of 33 dog breeds. Some areas correlated with guarding and some with companionship. Whippets, which are the best dogs, showed developed brain areas associated with sight and spatial movement. Cockapoos, which are also the best dogs, were not included in the study as they are a cross-breed, but poodles scored highly in terms of brain networks for olfaction and vision. All the dogs had highly developed areas of their brains for being very good boys or girls.

Now, while we can't include an emotion as inscrutable as love in this, the selection preferences applied by us have made dogs capable of certain desirable behaviours, such as loyalty, companionship and affection.

So, back to the question: do our dogs love us? Science is not just about collecting data and analysing it. Science is an ever-moving target. We lean into the truth, while simultaneously acknowledging that we don't ever reach it. Science is also an inherently social activity. It's as much about disagreement, discussion and argument as it is about data. A subject such as emotions, and the inner mental state of a dog, is opinion-rich yet data-poor, which means it is a fertile ground for debate.

And this is a topic about which your authors disagree with each other, which is a rare but happy occurrence.

Adam: Because love – as defined in any way – is necessarily a human concept and can only be expressed by humans, the answer has to be no, Jesse does not love me. Only humans are capable of love because it is a human condition. The feeling that Jesse has for me is a dog feeling, and therefore ineffable to me. Until he learns to speak, he can't describe to me how he feels. We have selectively bred dogs over thousands of years and millions of generations to be loyal, useful and effectively to mimic children, but much more obediently, and his relationship with me is based upon safety and protection, and provision of food, treats and scritching. Even though all these things nurture a behaviour that resembles love in so many ways, that is the limit of what we can say about Jesse's feelings for me. He is, predictably, a maniac, but one with such charm and adorableness that he's impossible not to love, unless you are a cat. But in attempting to train this beautiful, joyous idiot I employ a strict reward system; it's clear to me that the love we share is heavily based on the small, chicken-flavoured treats that now fill my pockets.

It is impossible to understand the inner experience of another human, but we agree through consensus the sense of tastes, or colours or emotions, or even what love feels like. We cannot establish a consensus with a dog, in the same way that we cannot know what is going on in his lovely little brain when he sniffs the balls of other dogs in the park. This is dog stuff, and whatever Jesse feels for me, that is dog stuff too.

Hannah: What an absolute load of poppycock. Of course Molly loves me. A child's relationship with their carer is also based on safety and protection, but it would be preposterous to propose that our love for our parents only engages once we develop the capacity for the language to express it.

Love is a two-way connection. It is biological vanity to define the action of loving by how it feels to experience it within a human body, rather than how it manifests itself outwardly, through the shared experience of loyalty, companionship and attachment. And by the latter metric, there is no doubt that dogs are entirely capable of love.

More importantly, even if Adam is right, this is a real-life situation akin to the philosophical thought experiment, known as the Chinese Room, in which a message, written in Chinese characters, is passed through a letter box into a sealed room. Inside the room is a person who composes meaningful responses, also in Chinese, and posts them back out. The conundrum is that with no way of looking inside the room, you cannot tell whether the person inside is fluent in Chinese or has no understanding and is simply googling how to respond so that the reply makes sense.

In the same way, even if Molly has no understanding of the messages of love that I give her, she gives the right response every time. A dog who possesses a brain that is capable of truly loving its owner is indistinguishable from one that is not. Therefore, we may as well believe that they do.

*

This conflict is in the file marked 'To be continued'. We cannot know what it is like to be another, be it cat, dog or human. What we agree on is that, though we disagree, it doesn't matter. Love is pretty difficult to pin down and define. Science's supremacy in understanding the universe is not best reflected in the study of love, and neither does it supply the best language to describe it. We are off out for a walk with our respective pets and will leave this one to painters, pop stars and foppish poets.

CHAPTER 9

THE UNIVERSE THROUGH A KEYHOLE

A few years ago, one of your authors found themselves sat on an Italian veranda, chatting to a former army dog trainer who was then working as the handler for two wolves owned by a local billionaire. As you might imagine, he had plenty of thrilling yarns, but it was his tale of dog-training gone wrong that was the most enlightening.

During any major conflict, vast sums of cash are often smuggled across borders and through checkpoints. This is a persistent and pernicious problem during wartime, as that money can go on to finance weapons and terrorist operations. Corking its flow presents a serious and important operational challenge for the military. Since dogs have an exceptional sense of smell, and serve as our obedient and long-standing colleagues, the army came up with the idea of stationing them at checkpoints and training them to sniff out any hidden stashes of cash.

The dogs, of course, rose to the challenge. In no time at all they were acing the training exercises, reliably signalling to their handlers when detecting bags or volunteers that had stashed cash of virtually any currency. The handlers took the money dogs out into the field to get them to uncover loot that was passing through checkpoints in conflict zones.

It was a disaster. The dogs were missing even the most obvious examples – people found afterwards to be carrying thick wads of illicit cash were waved through by the handlers, having been ignored by the pooches. The dogs were withdrawn from service, yet when they were taken back into the training facilities they again scored 100 per cent accuracy in sniffing out hidden dosh.

This mystery was solved when some bright spark spotted the one small difference between the cash in the training sessions and the cash in the war zone. To stop money going missing, the banknotes used in the tests had been wrapped in plastic before being placed in the pockets and bags of the testers. The army hadn't trained the dogs to smell money at all: they'd trained them to find clingfilm.

Dogs are often described as great smellers. This is undoubtedly true, as we have specifically developed some breeds to have noses that match the accuracy of expensive lab equipment, but they are of course more mobile and far more lovable. Dogs smell things in a very different way to humans. Where we breathe and smell through the same tube, they separate air for breathing from air for sniffing. Our olfactory bulbs sit in the relative open caverns of our nasal passages, and if you were to spread out that surface it would be about as big as the lid of a jam jar. Dogs have a labyrinth of bony passages called turbinates, and if you laid that out flat it would be very messy and you'd have a dead dog. But, more importantly, the surface area would be dozens of times larger (depending on the breed), more like a coffee table. This hugely increases the space for olfactory neurons to grab those odorous molecules from the air. In short, dogs smell much better than we do.

Humans, by contrast, may not be in the smell premiership with dogs, but we shouldn't be ashamed of our noses. In fact, we can do some of what dogs do, when pushed by scientists to perform. In 2006 a group of researchers set out to demonstrate that humans might also be capable of following a scent on the ground. They got a group of 32 human subjects to follow the trail left by some twine dipped in chocolate essence and dragged across a lawn for 10 metres, including one 45-degree bend. To make sure that they were indeed being led by their noses and nothing else, the subjects were blindfolded and wore gloves and earmuffs. They were then told to crawl along the ground for ~~maximum humiliation~~ the closest re-creation of the way in which a dog behaves, sniffing the grass as they went. The human sniffers fell short of the standard you might expect from your average bloodhound, but in most cases they were indeed able to follow the trail, including the bend. What's more, the more they practised, the faster and better they became.

Humans can distinguish between odours and flavours at the atomic level. Spearmint and caraway are the perfect example: the molecules that give them their distinctive flavours are identical, except for being mirror images of one another, and our noses can identify them. This is like being able to tell left from right simply by smelling a glove. Estimates vary, but credible calculations put the number of potential smells that humans can detect in the trillions.

And yet, even though we have the ability to detect the faintest of odours (especially when forced to by scientists), smell doesn't play a major part in our experience of the world. For a dog, however, the way it navigates reality is heavily dominated by its perception of a rich tapestry of smells. All animals perceive the universe in different ways. Scientists sometimes call this 'umwelt', from the German word that roughly translates as 'environment' or 'surroundings', but it's come to mean the more general and nebulous concept of the subjective universe. The objective universe does exist, of course, but umwelt allows us to recognize that even though we share our environment with dogs, our subjective worlds are simply not the same.

It's not just dogs. There are lots of animals whose existence is much more dependent on odour than ours, animals whose lives are utterly dominated by the smellscape that they perceive and we are oblivious to. Scent plays a vital role in the way that many creatures interact with the world and each other, notably in what are sometimes lovingly referred to as the four Fs of evolution: Feeding, Fighting, Fleeing and Reproduction.

The four Fs

Ants are masters at using scent to assist their feeding. You may wonder how they magically appear after seconds into every picnic on Earth, especially as they have no nose. How do they even smell? Via their antennae. As with most insects, their antennae are studded with olfactory detectors, and they swing their heads from side to side to pick up the trail of anything smelly on the ground or in the air. Smell is how they navigate the world. They are sensitive enough to detect the faintest whiff of sugars, oils and proteins, and can even differentiate between different types of coffee. Scouts will then lay down their own trail of hydrocarbons for their hive mates to follow and descend on your jam sandwiches. And they're not wasteful; once a food source has been finished, they'll lay down a signal to repel other ants from wasting their time looking for an empty larder.

When it comes to fighting, aggression or warning signs, cats, lions, elephants and a whole host of beasts – including some that live underwater – will mark their claimed territory with odours, as a means of telling others to back off. Male longfin squid compete for access to females, and when they see eggs on the seabed it means that a fertile female is nearby. There's a particular protein molecule on the surface of the eggs that turns a peacefully swimming male instantly into a rage-filled maniac who will attack any other males nearby by butting, grappling and generally beating the hell out of them.

As for fleeing, we discovered in the chapter on free will that rats pathologically avoid cat pee, unless they are pathologically infected with toxoplasmosis. Many animals have evolved an innate fear of certain odours from other animals that want to eat them. In 2001, researchers discovered a particular set of olfactory receptors that were stimulated by one specific molecule that is found in the urine of many predators, called 2-phenylethylamine. In a prime example of the glamour of science, David Ferrero and his colleagues collected the urine from 38 species in zoos around America, from lions, snow leopards and African wildcats to cows, giraffes and zebras. They found that the meat-eating predators produced far more

2-phenylethylamine than the herbivores, and rats would steer clear of this chemical on pain of death. This may be the reason why an animal that is a potential prey will know to give a wide berth to an animal which is a predator, even though they have never encountered one before.

And as for sex, well, smell is the most potent attractant for many animals. Sex pheromones can be mesmerizing for some. If you are a female boar, 5α-androst-16-en-3-one is one of the most exciting smells you can imagine, though she probably calls it something different. Male boars produce this hormone in their saliva, and when a female on heat smells it, its power is such that she will adopt a physical position called lordosis, a posture of sexual readiness, also known as the 'mating stance'.

Humans do none of these things. Not even whatever it is those boars are up to. We don't run away from predators because we can smell them. We don't mark our territories with our urine, at least not in polite society. And while Jesse, the whippet who featured heavily in the previous chapter, is capable of producing some of the most remarkable smells that either of us has ever had the misfortune to experience, even they are not bad enough to send us into a furious rage. You could argue that we are drawn to the smells of nice food wafting out of restaurants, but it's not quite the same biological imperative as ants have at a picnic. When it comes to sex, there's a whole industry out there that sells products claiming to attract sexual partners (almost exclusively for men to attract women, predictably), but there has never been any evidence supporting the idea that sex pheromones even exist in people. Certainly, bad-smelling people are generally less successful on a night out, but that's not the corollary of secreting a chemical to turn someone on.

We seem to have evolved out of our reliance on smell as a primary means of communicating with each other and with other animals. But this ability does appear to be something that our ancestors possessed in abundance. Olfaction is an ancient sense, and its underlying genetics contains clues to humankind's shortfalling in matters of the nose.

To smell or not to smell

When genes produce proteins that are of no particular use, they can mutate with impunity and become redundant, free to rust in the genome, like werds thet got ugnored by teh copyedytor. But by looking carefully, we can see that what were once working genes have lost their function. Nowhere is that more apparent than in the remnants of our rich-smelling ancestors. We have almost 900 bits of the genome that are related to smell, but half of them have rotted. The genetics of the ways in which our sense of smell has declined has resulted in differences between what people can and can't smell. 5α-androst-16-en-3-one, so beloved of horny female boars, is a prime example: to some people it's sickly-sweet, others simply lack the biological hardware to detect it at all.

Another example is asparagus. When you eat some delicious young asparagus, within half an hour your wee will honk. Your body will digest the asparagusic acid from the vegetable and convert it into various sulphur-rich compounds, which are excreted in your urine – methanethiol, dimethyl sulphide and a few others. To some people it smells chonky and heady, others can't smell it at all. That binary ability to smell some things or not is simply due to whether you have a working version of a specific gene, or it has mutated to become out of order.

Somewhere in all those genetic permutations between us, there are some extraordinary humans – people whose smelling sensitivity can have remarkable consequences.

How to smell a lethal disease

Joy Milne, a retired nurse from Perth in Scotland, had noticed a change in her husband's aroma. Les had developed a kind of indistinct musky odour to his skin. She accused him of not brushing his teeth or not showering properly. He was adamant that he was taking care of himself, so she let the matter lie. Six years later, Les was diagnosed with Parkinson's disease and subsequently died from that debilitating condition.

After the diagnosis, Les and Joy attended a Parkinson's support group together. As she sat in the room, Joy was confronted with the very same smell. She pretended to busy herself handing out cups of tea, using it as an excuse to sniff the other attendees; she discovered they whiffed just like her husband. She mentioned this to Parkinson's researcher Tilo Kunath from the University of Edinburgh. Though this seemed uncanny or even unlikely, he believed Joy and decided to test her. First, he gave her T-shirts that had been worn by six people who had Parkinson's and six people who did not, and asked her to identify which were which. Joy got 11 out of 12 right, but detected the distinctive smell on one T-shirt that belonged to a person without the disease. Eight months later, that person was also diagnosed with Parkinson's.

This remarkable feat became the basis for a whole research

programme that resulted in the discovery of a combination of aromas – including exotic-sounding molecules such as hippuric acid, eicosane and octadecanal – that are concentrated in the oily sebum secreted in the skin of people who are going to develop the disease. We don't yet know why this happens, and particularly why it happens before the onset of other telltale signs. Parkinson's is normally diagnosed via observation of a range of symptoms: tremors, slowing of movements and, ironically, anosmia – a loss of smell. With the work published in 2019 that started with Joy Milne's super-smelling, we may now have a new way of detecting this devastating disease months earlier than ever before.

What this glimpse into the maelstrom of olfaction shows is that the umwelt of smell doesn't just vary between species: there isn't a quintessential 'dog experience' of smell and a 'human' one. The basic biology of genetics means that humans' ability to engage in the world of smell is not just limited, it is unique to each of us. What you can smell is different from what we can smell. And what one of your authors can smell is different from what the other can. Add this to the heady cocktail of experience and the way in which a scent can trigger memories of joy or pain, and you get a sense of smell that is totally unique to each of us. Umwelt is inherently, absolutely and wholeheartedly personal.

Of all our senses, smell is perhaps the most emotive. How powerful is that moment when a whiff of something instantly transports you back to your childhood, or a holiday years ago? Sizzling bacon in the morning, the pages of an old book, baking bread, vinegar on chips. Is there a more potent sensation than the smell of a baby's scalp, or the scent of a lover's skin? It has the power to trigger the most evocative memories. For Adam:

THE MOST USELESS SUPERPOWER

Adam writes: while we were writing this book, a new infectious virus turned the world upside down. Millions have been infected, and millions have died. One of the common symptoms of Covid-19 is what is called anosmia, a radical loss of the sense of smell. About half of Covid patients develop anosmia, and as far as we can tell (to date) around 90 per cent of them recover after about a month. We don't yet understand why this happens. One theory, put forward by the biologist and smell expert Matthew Cobb, is that the virus knocks out a protein that helps chaperone volatile molecules from the air to the olfactory bulb cells, but, by his own admission, this is just an educated guess.

I had my own pretty close shave with Covid in March 2020 – not fun and I do not recommend it at all – but, unlike most Covid patients, I had a different smell-related symptom, which seems to affect a very small proportion of people – hyperosmia, whereby the sense of smell is heightened, also for unknown reasons. Basically, I could smell everything. I correctly identified the hand cream that my daughter was applying two floors away in our house. I could smell the shower gel my son was using behind two closed doors. I could even smell the gentle incense sticks that were being burned in the corner shop across the road. Superheroes in comics often acquire their powers through adversity – the unkillable Deadpool via a normally incurable cancer, Daredevil from having been blinded as a child. The ability to smell hand cream is what I got from Covid, and, frankly, I feel a bit short-changed.

the rich smell of sugar beet immediately transports me back to my youth in Suffolk playing rugby in fields near the factories where sugar was being refined. Hannah, knowing that smell can trigger memories, bought a posh scented candle and lit it every day while on her honeymoon, in the hope that on her return she could for evermore ignite those same warm feelings at the flick of a match. Unfortunately, they discontinued the candle and she has never been able to find one for love nor money. Her marriage fell through and now she lives with an army dog trainer and two wolves.*

For some, smells evoke powerful post-traumatic stress, as memories flood back of times in wars, or when they witnessed injury or pain. Our (limited) understanding of this phenomenon relies on the discovery, in a part of the brain called the hippocampus, of a set of neurons called place cells, which are active when we recall specific locations. These cells take cues from many sensory inputs, including smells, and are subsequently triggered again when we re-encounter whiffs from our past.

We wander through the world with little idea of how others experience it. And smell isn't even our dominant sense. The difference between our own perception and that of others becomes even more profound when we switch from our noses to our eyes. In vision, too, reality is not what it seems.

Time for your eye test

To do this you'll need a friend and a brightly coloured object, such as a marker pen. Don't allow the friend to see the object.

---

* This, for the avoidance of doubt, is what's commonly referred to as a joke.

Tell them to look straight ahead while you hold the object to the side, about a metre out from one ear. As they stare forwards, slowly move the object into their field of vision, tracing the arc of a semicircle around the front of their head, and ask them to tell you when they see something. As soon as they do, stop, and wiggle the object. They should be able to see the object wiggling but not what it is, and, more importantly, not what colour it is.

The main function of an eye is to capture photons, and in mammals that happens in the retina – three layers of neurons which are basically an extension of the brain. The work of collecting photons is done by cells called photoreceptors, which come in two types: rods and cones. The job of rods is to detect movement, and in humans they are stationed around the edge of the retina. They can't perceive colours, and don't give you high resolution.*

When you pause that object at the periphery of your friend's vision, it is only the rods that are firing. Which means that your subject can only see the existence and movement of the object, but not its colour or detail.

As you keep moving it along the arc, a point will come when the cones kick in. Cones are the colour catchers. Because they are densely packed in the centre of the eye, they are the cells that give you the detail of an image. You need the cones to be firing for the identity of the object to be revealed.

If you trusted your brain, you would think that you had full colour vision. But you don't. That wiggly demonstration shows that you're seeing in black and white at the sides. Perhaps you

---

* We mentioned rods in Chapter 5, as one of the reasons we think we see ghosts, because the rods are tuned to work best at twilight and to detect monochromatic movement out of the corner of our vision.

never noticed. Normal people don't tend to spend that much time thinking about looking,* but waving marker pens around shows that there are physical limits to what our eyes can do, and that restricts what and how we can see. Vision is not an objective snapshot of reality.

It gets worse. The densest concentration of cones is in the fovea, where our vision is sharpest, and most useful for doing what you are doing right now. The nerve fibres that flow out of the photoreceptors bundle up into the optic nerve, which connects the eye to the brain. But where the optic nerve exits the eye there are no photoreceptors at all and we have a blind spot, a place of retinal darkness called the scotoma. We can show you where this is by making one of us disappear.

First, look at the lovely pictures of your authors below from about 2 feet away, your nose in between our heads. Then close your left eye and focus your right eye on Hannah's face.

Now slowly move the book closer, all the while staring at Hannah.

* Adam does, though.

At some point, probably about a foot away, Adam's face will vanish from your sight. The photons from Adam's face are still bounding into your eye, but they're hitting the blind spot and therefore not being turned into electrical signals. Your brain is aware of this hole in your photoreceptors, so it fills it in with the visual information that it *can* process, from the areas around Adam's face, which is whiteness, and he briefly ceases to exist.

You can reverse this by closing your right eye and focusing on Adam to make Hannah disappear, if you prefer. Although quite why you would want to is anyone's guess.

Sight dominates our sensual experience of the world. Of course, some people are born without sight, or become blind during their lives; but for most people the universe is bathed in light, and we take those photons into the darkness of our skulls and construct a rich, colourful view of reality.

Do you see what I see?

Cones, the colour catchers, come in three flavours. Inside each cone are molecules called opsins that hoover up photons. These molecules are exquisitely tuned to pick up light in one of three specific wavelengths. Roughly speaking, that tuning is honed to recognize light in the short, medium and long wavelengths of the visible spectrum, but it's easier to think of your cones as blue, green and red – that is, the primary colours.

Depending on their genetic make-up, it is possible for people to lack any one of those three cone types, leaving them unable to distinguish between certain colours. Most of these conditions are extremely rare, except for red-green colour-blindness, which appears in around 8 per cent of men (and only 0.5 per cent of women).* It mostly arises when there is some fault in the genes that encode for either the red or the green cones. Where that happens, a person is left with only two functional cones: blue and whatever else is left. When a photon from the red-green end of the spectrum then enters the eye, the remaining cone has to pick up the slack. If, for example, your green opsin is broken, then the remaining red opsin has the job of sucking up both green and red photons, despite being unable to distinguish between them.† A person with this condition can't see the difference in colour between a red cricket ball and green grass, or between the red and green of a traffic light (which is why they are designed with bulbs in different places rather than one bulb which changes colour).

Still, despite what must be dramatic differences in the experience of vision between people who have three working cones and those who don't, a large proportion of people with defects in their colour vision are entirely unaware of their condition. We are so lacking in a universal language with which to describe our visual umwelt that many of you reading this could well be wandering around with a world-distorting genetic irregularity without even knowing it.

* The genes that encode green or red cones are situated on the X chromosome. That means that if you were born male, and are unlucky enough to have inherited a wonky green opsin gene (for example) from a parent, you have nothing else to fall back on. Females, by contrast, have two X chromosomes, which gifts them a second roll of the dice wherever the faulty genes are handed down on those stretches of DNA.
† Red and green opsins overlap in their absorption spectra, meaning that they can pick up both types of photon. But red-green colour-blindness is only one of the many forms of colour vision disorders. The blue opsin can also break, meaning that people can't see colours in the blue range. We are simplifying here because genetics is ridiculously complicated and Hannah is having a bit of a wobble.

Or maybe you've got a genetic superpower instead. Some women may have acquired another cone altogether, which means that they might be tetrachromats – that is, they see in four primary colours. This is a new phenomenon (one that we don't fully understand yet), but it seems that about one in eight women have a novel bit of DNA – another version of an opsin gene on their X chromosome. We know that some people have the ability to see subtle variations in shades of colour where the rest of us see monotones, and this might well be why. It could be that some of us have the genuine ability to distinguish thousands more colours, just without the language to name them. Women: if you find yourself having polite disagreements about shades of colour, or if you think you see rich, multicoloured greens where others see murk, you might just have reached the next stage of human evolution.

We know that our experience of the world varies and is limited by our fundamental biology. We know that some of us see in more colours than others. And we know that perception happens in the darkness of our skulls, where the information from those photons is processed and reconstructed into our lived realities. What we don't know – and have no real way of testing – is whether our brains have constructed the same internal experience in response to the photons that flood into our eyeballs.

We may all agree that a cricket ball is red. Red is the word we've settled on to describe the wavelength of light that bounces off its surface and is captured by our retina. But is your red the same as someone else's? If, somehow, you could transport your mind into theirs, to experience their view for a day, would it differ dramatically from your own?

The precise experience of colour – or taste, or smell,

or touch – within each person's mind is scientifically unmeasurable, unknowable and ineffable. This is one of the hardest problems in science and philosophy, so much so that, in an act of stunning originality, scientists and philosophers refer to it as 'the hard problem'. Yet whenever science tries to codify our perceptions of reality, all it finds are differences. There is biological variation in our perception of colour, our ability to smell, our sensitivity to taste. No one else has experienced – or will ever experience – the world in quite the same way as you, which might just help explain the huge differences in the things that people like, the colours they think look good and the smells they prefer to avoid.

Beyond the visible

It's a mistake to think that one person's experience of the world is the same as someone else's. And it's equally wrong to assume that our species' world view is the same as that of any of the billions of other creatures with whom we share this planet. The limits of our perception are set by our hardware, and our hardware has evolved for the world we live in.

Nowhere is this more apparent than with vision. The rainbow we see is made of the same stuff that bathes the universe – photons of different energies behaving as waves of different lengths. Our eyes perceive those colours (and agree they have colour to them) because that's what our biological hardware can detect – but there's nothing special about those particular packets of light. The spectrum of electromagnetic waves extends way beyond the ones that we can detect, from the X-rays that we use to view our bones, microwaves that heat up our food, cosmic rays that pose such danger to astronauts, radio waves that fill our homes with music and politics and curious cases of scientific investigations. They are all made of the same stuff.

It's a big old spectrum too. Gamma rays are the shortest, with a wavelength of 1 picometre (that is one-billionth of a millimetre). At the other end of the scale there are extremely low-frequency radio waves, which have a wavelength of around 100,000 kilometres.

Our cones can detect only the narrowest sliver of the electromagnetic spectrum: waves that are somewhere between 370 nanometres – deep violet – and 700nm – ruby-red. We might be able to see a little in the ultraviolet range, but our lenses block out UV light altogether. So our hardware literally limits the range of what we can see to a tiny fraction of the light that fills the universe. If the entirety of this book, with 304 pages and 350,767 characters, were the EM spectrum, we would be able to actually read less than a sentence of it. The only bit visible to us would be a around 12 letters, which amounts to: not a whole lot. Other animals are not quite so limited. Many flying insects see in the ultraviolet. And flowers, which need insects to pollinate them, know this well; their pretty petals often have ultraviolet runways pointing down towards their juicy nectar, and the all-important sexual organs. Honeybees see in the UV very well, though not in the red part of the spectrum like we do. Their eyes, which have a higher flicker threshold and can also perceive iridescence, are superbly evolved to spot flowers while hurtling along on the wing. Just as long as they're not red.

Among the insects, though, it is the butterflies that seem to enjoy the richest colours. This is not simply because they happily detect light in the UV range, but also, while we humans only have a measly three primary colours (or possibly four), many butterflies have nine or ten, and in the case of the common bluebottle butterfly – which is a butterfly and not a bluebottle – it's 15. Their worlds are bathed in colour in a way we cannot imagine. Think of Dorothy when she steps into Oz, out of the dreary black and white world of Kansas and into the technicolour of munchkins, the Wicked Witch of the West and the yellow brick road. That would be us, if only we could see what a butterfly sees.

There aren't many mammals that can see in UV, but reindeer can, and we have discovered this by knocking them out. Even when an animal is anaesthetized you can test its retinal range, and in 2011 a group of researchers did just that, by shining lights into the eyes of 18 reindeer. With lights in the UV range, their retinal neurons fired. This makes sense when you consider the world the reindeer inhabits. UV is mostly absorbed by the ground, but when it's covered in ice or snow almost all of it is reflected. However, lichen (an important food for reindeer), and urine (an important signal for fighting and fornicating) appear dark against the glare of the UV bouncing off the snow. What looks like a blanket of pure white snow to you or me is stained by the telltale signs of food and sex to the reindeer.

More recently, it was discovered that the platypus emits multicoloured UV light. It is unknown if they can see in this spectrum, but we figure this is all very much in keeping with this utterly bizarre creature's other peculiarities. If you're going to be an egg-laying mammal with poisonous spurs and a beak that can detect electricity, then why not also have disco glow-fur?

MULTICOLOURED RULERS OF THE DEEP

Being able to see UV, though, is nothing compared to the true seers of vision. The queens and kings of the light live in the murk of the deep sea: the mighty mantis shrimps. There are many species of these mini-lobsters, and we think they deserve much more adoration than some of the dullards of the oceans.

These crustaceans are as colourful as a rainbow prawn at a Pride carnival. The image above, glorious though it most certainly is, cannot do justice to the peacock mantis shrimp. This is not the fault of our talented illustrator, Professor Alice Roberts. It is simply that humans are not even capable of viewing this beastie in all its magnificence. The peacock mantis shrimp can see light with wavelengths between 330 and 700nm, which is a wider range than most animals have, us included. But – and this is where shrimpy's true superpower lies – they also hold the record for different types of colour photoreceptors – 16, meaning an unimaginable ultra-technicolour array of 16 primary colours.

When it comes to visual resolution, that is determined by the number of units that make up the compound eye. Honeybees have around 150 per eye. The fruit fly has about 700 per eye. The mantis shrimp has 10,000. That's like the difference between playing *Space Invaders* on an Atari 2600 in the 1980s and *Assassin's Creed* on the PlayStation 5 in super HD on a 4K monitor.

This ultra-vision is all a bit puzzling, though. Mantis shrimps typically live almost a mile down at the bottom of the ocean, an inky place where few colours can be seen anyway. We don't know why they have such amazing colour vision – whatever the reason, it must be doing something important for the mantis shrimp. We just don't know what. It is not our umwelt. There is a whole world out there, or down in the oceans, or right in front of our noses that we are simply incapable of detecting.

The director's cut

At the very beginning of this book, we asked you to close your eyes. Now that we are near the end, we want you to swivel them. Get out your phone and flip the video camera so you can see your own face, about 20 centimetres from your nose, and press record. Now we want you to look at your own eyes on the screen. Look at your left eye, then switch to your right. Go back and forth a few times.

When you play that recording back, you will be able to see that your eyes move back and forth. It should not come as much of a surprise that eyeballs move – there are seven muscles in each eye, which would be pretty pointless if they didn't.

Now we want you to find a mirror and do the same. Up close, about 20 centimetres away, look at each eye one after the other, back and forth. You should find that there really isn't much to see here. Nothing visibly moves. We realize how unusual it is to do an experiment where the correct outcome is nothing at all, but stick with us for a moment. There's something quite extraordinary in that absence of anything to see. Obviously, your eyes have moved, otherwise you wouldn't be able to focus on each individual eye. But you can't actually see them moving. Don't panic, you do not need to consult an ophthalmologist, there's nothing wrong here, it's just that what you are trying to do is impossible. It is literally not possible to see your eyes moving with your own eyes.

What is this sorcery? Well, it is simply your mind's way of coping with the sheer volume of reality. On the filmed version of your swivelling eyes, you might notice that the movement of your eyeballs is not smooth but jerky, like a spasmodic second hand on a ticking clock. This might seem a bit strange, because when you pan your eyes across a beautiful vista, or a particularly long sausage, the image you perceive is continuous and smooth. And yet what your eyes are actually doing is juddering across the scene (or sausage).

These little flitting eye movements have a name: they are called saccades, and they are pretty superheroic. They are among the fastest muscular movements that humans can make – saccades dart across 500 degrees of visual angle per second (think: two degrees is about the width of your thumb if held at arm's length) – and we do up to four of them every second. When we look at a face, or a painting, or anything that we effectively perceive as one image, what our eyes are actually doing is scooting around and sampling that image in order to build up the whole picture. You are doing it right now. Reading is an activity where our saccades are in furious, juddery action, and though you might not know it, your eyes are darting and stopping across this line as you ingest these words in a way that is totally beyond your control* – the saccades happen as fast as they are able – but your brain is assembling the images into a coherent, readable and, dare we say it, very well constructed sentence.

Our eyes are not like the megapixel camera on your phone. Digital cameras have an array of sensors that pick up the bits of information from wherever the lens is pointing and convert a version of that view into an image. Our eyes don't work that way.

If you want a sharp, high-resolution image, your eye has to get photons on to the fovea – the tiny dent at the back of the retina with the densest concentration of cones, and hence the point at which we have the sharpest vision. Therefore, our eyeballs zip and zap all over the place. Our eyes don't take a smooth video of the world in front of us; they take continual snapshots, and our brains put the pieces together.

The reason why we can't see our eyes moving with our own eyes is because our brains edit out the bits between the saccades – a process called saccadic suppression. Without it, we'd look at an object and it would be a blurry mess. What we perceive as vision is the director's cut of a film, with your brain

---

* It's not totally beyond your control. You could close your eyes. Or pull a woolly hat over your eyes. Or put the book down and go for a sandwich. But you know what we mean.

as the director, seamlessly stitching together the raw footage to make a coherent reality.

Perception is the brain's best guess at what the world actually looks like. Immense though the computing power of that fleshy mass sitting in the darkness of our skulls is, if we were to take in all the information in front of our eyes, our brains would surely explode.* Instead, our eyes sample bits and pieces of the world, and we fill in the blanks in our heads.

This fact is fundamental to the way that cinema works. A film is typically 24 static images run together every second, which our brain sees as continuous fluid movement – that's why it's called a movie. The illusion of movement actually happens at more like 16 frames per second. At that speed, a film projection is indistinguishable from the real world, at least to us. It was the introduction of sound that set the standard of 24 frames per second with *The Jazz Singer* in 1927, the first film to have synchronized dialogue. The company that made the sound-recording system decided to use a motor to power both the sound disc and the film reel so they would not fall out of sync – earlier attempts had both sound and vision powered off different motors. The new system set the frame rate at 24, for no particular reason. The rest is movie history.

This works for humans but is certainly not universal: the frame rate used in cinema would make no sense to the visual system of, say, a pigeon.

It's the pigeon's own frame rate that gives rise to its distinctive bobbing. Except it's not bobbing. It's a form of image stabilization. If you were to put a pigeon on a treadmill, you'd

* They wouldn't explode. But it would be a hot, fuzzy mess, like drunkenly waving a camera about a party. Do try this at home.

MOTION BLINDNESS

It's often the case that we learn about how something works when it is broken. There is a very rare neurological disorder called inconspicuous akinetopsia which highlights this very phenomenon. It's a kind of motion blindness, whereby people cannot see the illusion of the moving image. Instead they see the individual still frames, a bit like watching people dance under a strobe light at a disco. Scientists don't really know why this disorder occurs, but it involves a part of the brain called the V5 segment of the occipital lobe, round the back left of your head. There, motion is processed from visual cues. People with inconspicuous akinetopsia suffer from disruption in this part of the brain – sometimes brought on by prescription drugs. Patients report finding pouring cups of tea difficult, because they see a frozen image of the cup half full, before it strobes instantaneously to the cup running over without any movement in between. This scenario is considerably more dangerous when you replace tea with a moving car.

see that its head stayed locked for up to 20 milliseconds while its body moved underneath it. We know this, because this exact experiment was done in the 1970s. Pigeons' heads don't actually bob at all; they keep still for as long as possible – taking time to ingest the visual information – before jerking forwards for the next snapshot. It's the same with hummingbirds and kestrels, which hover to keep their heads as still as possible, and long-necked birds such as geese, whose head movements balance out the forceful downward thrust of their beating wings. So if you took a pigeon or a goose or a kingfisher with you to the cinema, it would be incapable of enjoying the film. The flashing images would be incomprehensible. Likewise, the great pigeon cinematographers would be wasted on us. And not just because all the storylines would revolve around pooing on statues.

The real guide to reality

It's not that pigeons are stupid. Well, they are, but that's not the point. This gulf between how we see the world and how a pigeon sees the same world reveals something fundamental about our relationship with reality, and how we understand our place in the cosmos.

Our eyes powerfully illustrate the fact that our experience is a heavily edited version of reality. Evolution has found a way for us to harvest, process and interpret elementary packets of light in the dark cavities of our skulls. Our minds navigate the many constraints of anatomy to make it work – frame rates, blind spots, faulty cones, colourless peripheral vision. We don't even notice the limits of our eyes as we construct our subjective world view in our heads.

Like all creatures on Earth, our bodies are carefully tuned to ensure our continued survival. But it would be a pointless waste of ego to think that they make us capable of experiencing reality as it really is. We are each locked into our own umwelt, profoundly limited by our senses, constrained by our biology, shackled by the inescapable bounds of our evolutionary history. We're hopelessly tethered to what we can uncover while stuck on (or perhaps near) this planet, a speck of dust in the vastness of the cosmos. We see only the merest sliver of reality. We're peering at the universe through a keyhole.

Yet thanks to science and maths, and insatiable curiosity, we know that there is so much more than we see, and hear, and smell, and touch, or can even imagine. Our brains come pre-installed with a whole battery of glitches and errors, which means we have to fight our prejudices, preconceptions and biases. But we also come pre-installed with the burning desire to do so. The very fact that we can recognize that our perception is limited and skewed, and human, is precisely what gives us the ability to unskew our faulty intuitions and go beyond those limits.

This is our glorious purpose. We *can* see the totality of the electromagnetic spectrum, from X-rays to the Hawking Radiation leaking out of otherwise invisible black holes. We may not be able to reliably perceive time, but we do know that we can't, and correct for that by building clocks that don't lose so much as a second throughout the entire duration of the universe. We may not have the olfactory sensitivity of dogs, but we can tell you with exquisite accuracy – because astrophysicists have identified the presence of ethyl formate in the heart of our galaxy – that the Milky Way smells like rum and raspberries.

Look how far we have come. We've exceeded our programming and reached way beyond our grasp, into the depths of our cells, the crevices of our minds, the structure of atoms and the fabric of the universe. In the last few thousand years we have developed science, the only tool capable of seeing the world as it really is rather than as we perceive it to be. It is not without flaw, but only science can ever take us beyond our biological limits from the subjective to a genuinely objective view. Science is now – and will always be – the only way to compose the ultimate guide to everything.

ACKNOWLEDGEMENTS

The following people helped us in small and large ways, and we're very grateful to them all:

Will Storr, Sharon Richardson, Stuart Taplin, Natalie Haynes, Thony Christie, Michelle Martin, Matthew Cobb, Andrew Pontzen, Leon Lobo, Robert Matthews, Julia Shaw, Lisa Feldman Barrett, Anil Seth, Rebecca Dumbell, Louisa Preston, Stephen Fry, Cori Phillips and Alice Roberts.

Special thanks, as always, to Georgia, Beatrice, Jake, Juno and Jesse for keeping us entertained and fed for a long summer, and to Phil, Edie, Ivy and Molly for their unwavering support – you are the perfect antidotes to writing, in both restorative and often quite distracting ways.

Thanks, too, to Will Francis and Clare Conrad at Janklow and Nesbit, and Susanna Wadeson at Transworld for her endless patience as the deadlines whooshed past us. We got there in the end.

Most of all, thank you to Adam, for being the Morecambe to my Wise (or maybe more the Richard to my Judy) and for your immense generosity, in both wisdom and friendship. And to Hannah, for standing by me when I was in trouble, and for making me laugh for the last five years.

PICTURE CREDITS

Illustrations on pages 20, 50, 64, 97, 153, 205, 270 by Professor Alice Roberts.

Illustrations on pages 27, 58, 83, 114 by Julia Lloyd.

Photographs on page 222 by Heritage Art / Heritage Images via Getty Images.

Author photographs on page 263 by Stuart Simpson / Penguin Books.

All other images are in the public domain.

REFERENCES

Here is a selection of references for specific papers that we cite in this book, and some other relevant studies, articles or just things we like that you might find interesting.

Introduction

The development of object permanence in babies is a fascinating and extensive field of research without consensus. A good place to start is Jean Piaget's theory of cognitive development.

Dung beetles wearing hats can't navigate:
Dung beetles use the Milky Way for orientation
https://doi.org/10.1016/j.cub.2012.12.034

It's parents, not children, who behave badly at parties, and sugar has nothing to do with it:
The effect of sugar on behavior or cognition in children:
a meta-analysis
>https://doi.org/10.1001/jama.1995.03530200053037

Chapter 1: Endless Possibilities

We urge you to go and have a play around in Jonathan Basile's version of Borges' total library, available at:
>https://libraryofbabel.info/

The arguably unnecessary study that showed that six Sulawesi macaques with a typewriter can't in fact write Shakespeare, but did use the keyboard as a toilet. It is available to buy for £25, though the complete text can be found here in its original form:
Notes towards the complete works of Shakespeare
By Elmo, Gum, Heather, Holly, Mistletoe & Rowan
>https://archive.org/details/
>NotesTowardsTheCompleteWorksOfShakespeare

Chapter 2: Life, the Universe and Everything

Images of potential cryovolcanoes on Pluto (as if the fact we have images of Pluto isn't interesting enough):
>https://www.nasa.gov/feature/possible-ice-volcano-on-pluto-has-the-wright-stuff

The September 2020 paper describing the presence of phosphine on Venus, which some enthusiastically interpreted as being a biosignature . . .
Phosphine gas in the cloud decks of Venus
>https://doi.org/10.1038/s41550-020-1174-4

. . . and the paper that said the opposite. Note: we consider this process to be good science.
No phosphine in the atmosphere of Venus
    https://arXiv.org/abs/2010.14305v2

Pakicetus, the dog-sized whale ancestor from Pakistan who got back in the sea:
New middle Eocene archaeocetes (Cetacea: Mammalia) from the Kuldana Formation of northern Pakistan
    https://doi.org/10.1671/039.029.0423

Hitchhiking barnacles on the bellies of ancient whales:
Isotopes from fossil coronulid barnacle shells record evidence of migration in multiple Pleistocene whale populations
    https://doi.org/10.1073/pnas.1808759116

Argentinosaurus walking:
March of the Titans: the locomotor capabilities of sauropod dinosaurs
    https://doi.org/10.1371/journal.pone.0078733
    https://www.manchester.ac.uk/discover/news/scientists-
    digitally-reconstruct-giant-steps-taken-by-dinosaurs/

The biomechanics of ants' neck strength:
The exoskeletal structure and tensile loading behavior of an ant neck joint
    https://doi.org/10.1016/j.jbiomech.2013.10.053

From the amazing journal *Superhero Science and Technology*:
Ant-Man and the wasp: Microscale respiration and microfluidic technology
    https://doi.org/10.24413/sst.2018.1.2474

The universal law of urination:
Duration of urination does not change with body size
    https://doi.org/10.1073/pnas.1402289111

Chapter 3: The Perfect Circle

The ongoing issue of astronaut eyeballs:
https://www.nasa.gov/mission_pages/station/research/news/iss-20-evolution-of-vision-research

Just how round IS the Sun?
https://doi.org/10.1111/j.1468-4004.2012.53504_2.x

The most perfect balls ever created:
Gravity Probe B: Final results of a space experiment to test General Relativity
https://doi.org/10.1103/PhysRevLett.106.221101

Chapter 4: Rock of Ages

The main reference for the first part of this chapter is the Bible. We're not sure how to cite it: various revised editions are available, and the authorship is unclear.

A really very old sponge:
Whole-ocean changes in silica and Ge/Si ratios during the last deglacial deduced from long-lived giant glass sponges
https://doi.org/10.1002/2017GL073897

Prehistoric baby bottles:
Milk of ruminants in ceramic baby bottles from prehistoric child graves
https://doi.org/10.1038/s41586-019-1572-x

Chapter 5: A Brief History of Time

The epic slowing of the Earth's orbit as told by old corals:
Proterozoic Milankovitch cycles and the history of the solar system
https://doi.org/10.1073/pnas.1717689115

What time is it?
www.bipm.org

The analysis which partially addresses the question of whether
President Trump was doing a poo while pushing out his most batshit
Twitter effluence:
Twitter as a means to study temporal behaviour
https://doi.org/10.1016/j.cub.2017.08.005

The most recent experiment on living in a cave:
https://deeptime.fr/en/

Time perception and schizophrenia
https://doi.org/10.2466%2Fpms.1977.44.2.436

Time's subjective expansion for an expanding oddball:
Attention and the subjective expansion of time
https://doi.org/10.3758/BF03196844

Time flies when you're eating cake:
*Time flies when you're having approach-motivated fun: Effects of
motivational intensity on time perception*
https://doi.org/10.1177%2F0956797611435817

Time does not fly when you throw someone off a building:
Does time really slow down during a frightening event?
https://doi.org/10.1371/journal.pone.0001295

Chapter 6: Live Free

The hypnotic mind-control zombification hexes seem so unreal
and utterly unbelievable, we thought we ought to share the original
research which describes their absurd behaviours:

The emerald cockroach wasp:
Direct injection of venom by a predatory wasp into a cockroach brain
https://doi.org/10.1002/neu.10238

The parasitic disco snail worm:
Do Leucochloridium *sporocysts manipulate the behaviour of their snail hosts?*
https://doi.org/10.1111/jzo.12094

The gordian worm:
Water-seeking behavior in worm-infected crickets and reversibility of parasitic manipulation
https://doi.org/10.1093/beheco/arq215

Crab hacker barnacles:
The selective advantage of host feminization: A case study of the green crab Carcinus maenas *and the parasitic barnacle* Sacculina carcini
https://doi.org/10.1007/s00227-012-1988-4

Ant mind-control zombie fungus:
Evaluating the tradeoffs of a generalist parasitoid fungus, Ophiocordyceps unilateralis, *on different sympatric ant hosts*
https://doi.org/10.1038/s41598-020-63400-1

Toxoplasmosis and humans:
Effects of Toxoplasma *on human behavior*
https://doi.org/10.1093/schbul/sbl074

People with flu party more:
Change in human social behavior in response to a common vaccine
https://doi.org/10.1016/j.annepidem.2010.06.014

2018 study on criminal behaviour after tumour or injury:
Lesion network localization of criminal behavior
https://doi.org/10.1073/pnas.1706587115

Benjamin Libet's original studies on readiness potential:
Unconscious cerebral initiative and the role of conscious will in voluntary action
> https://doi.org/10.1017/S0140525X00044903

Bereitschaftspotential in monkeys:
Bereitschaftspotential in a simple movement or in a motor sequence starting with the same simple movement
> https://doi.org/10.1016/0168-5597(91)90006-J

A coin-flipping machine:
Dynamical bias in the coin toss
> https://statweb.stanford.edu/~susan/papers/headswithJ.pdf

Tracking fruit flies' flight in a sensory deprivation drum:
Order in spontaneous behavior
> https://doi.org/10.1371/journal.pone.0000443

Chapter 7: The Magic Orchid

We're DOOOOMED! The Armageddon survey:
One in seven (14%) global citizens believe end of the world is coming in their lifetime
> https://www.ipsos.com/sites/default/files/news_and_polls/2012-05/5610rev.pdf

When Prophecy Fails is the classic investigation by Leon Festinger, Henry Riecken and Stanley Schachter of Dorothy Martin, the planet Clarion, and the cult of the Seekers.

Tom Bartlett's interviews with Harold Camping's devotees:
> https://religiondispatches.org/a-year-after-the-non-apocalypse-where-are-they-now/

Belief perseverance vs calculator:
Electronic bullies
https://doi.org/10.1080/07366988309450310

Belief perseverance vs equation:
Experimental studies of belief dependence of observations and of resistance to conceptual change
https://doi.org/10.1207%2Fs1532690xci0902_1

Bert Forer's classic study of how horoscopes work, and how we believe generic statements relate directly to us and us alone:
The fallacy of personal validation: a classroom demonstration of gullibility
https://doi.org/10.1037/h0059240

Studying aggressive ants:
Confirmation bias in studies of nestmate recognition: a cautionary note for research into the behaviour of animals
https://doi.org/10.1371/journal.pone.0053548

Chapter 8: Does My Dog Love Me?

For more on the adventures of Ada Lovelace and Charles Babbage, read Sydney Padua's excellent book *The Thrilling Adventures of Lovelace and Babbage*.

The Moon really was bright when Mary Shelley wrote *Frankenstein*:
The Moon and the origin of Frankenstein
https://digital.library.txstate.edu/handle/10877/4177

The faces of sightless judo fighters:
Spontaneous facial expressions of emotion of congenitally and noncongenitally blind individuals
https://doi.org/10.1037/a0014037

Surprise! It's Kafka:
Facial expressions in response to a highly surprising event exceeding the field of vision: a test of Darwin's theory of surprise
   https://doi.org/10.1016/j.evolhumbehav.2012.04.003

Read more about the problematic pseudoscience of emotion recognition in AI here:
   https://ainowinstitute.org/AI_Now_2019_Report.pdf

Meta review of Ekman faces:
Emotional expressions reconsidered: challenges to inferring emotion from human facial movements
   https://doi.org/10.1177%2F1529100619832930

The attempt (and failure) to replicate Ekman's study of Papua New Guineans' response to photos of sad-face actors:
Reading emotions from faces in two indigenous societies
   https://doi.org/10.1037/xge0000172

And more broadly, on the subject of emotions, you'll be hard pressed to find a better guide than:
How Emotions Are Made by Lisa Feldman Barrett.

Restaurant Row and rats with regret:
Behavioral and neurophysiological correlates of regret in rat decision-making on a neuroeconomic task
   https://doi.org/10.1038/nn.3740

Male flies get drunk when they crash and burn with females:
Sexual deprivation increases ethanol intake in Drosophilia
   https://doi.org/10.1126/science.1215932

Dogs and eyebrows:
Evolution of facial muscle anatomy in dogs
   https://doi.org/10.1073/pnas.1820653116

(or just do an image search of 'dog eyebrows' and lose yourself in the *joy of it all*)

The different brains of various good boys and girls:
Significant neuroanatomical variation among domestic dog breeds
https://doi.org/10.1523/JNEUROSCI.0303-19.2019

Chapter 9: The Universe Through a Keyhole

Humans following a scent:
Mechanisms of scent-tracking in humans
https://doi.org/10.1038/nn1819

The subjective universe of umwelt:
Jakob von Uexküll: the concept of Umwelt *and its potentials for an anthropology beyond the human*
https://doi.org/10.1080/00141844.2019.1606841

Predator wee:
Detection and avoidance of a carnivore odor by prey
https://doi.org/10.1073/pnas.1103317108

The woman who could smell Parkinson's:
Discovery of volatile biomarkers of Parkinson's disease from sebum
https://doi.org/10.1021/acscentsci.8b00879

Ultraviolet reindeer:
Arctic reindeer extend their visual range into the ultraviolet
https://doi.org/10.1242/jeb.053553

Glowing platypuses:
Biofluorescence in the platypus (Ornithorhynchus anatinus)
https://doi.org/10.1515/mammalia-2020-0027

Pigeon bobbing:
The optokinetic basis of head-bobbing in the pigeon
https://doi.org/10.1242/jeb.74.1.187

The smell of space:
Increased complexity in interstellar chemistry: detection and chemical
modelling of ethyl formate and n-propyl cyanide in Sagittarius B2(N)
https://doi.org/10.1051/0004-6361/200811550

INDEX

Hannah Fry is an Associate Professor in the Mathematics of Cities from University College London. She is also the author of *The Mathematics of Love*, *The Indisputable Existence of Santa Claus* and *Hello World* and regularly writes for *The New Yorker*. In her day job she uses mathematical models to study patterns in human behaviour, and has worked with governments, police forces, health analysts and supermarkets. Her TED talks have amassed millions of views and she has fronted television documentaries for the BBC and PBS. With Adam she co-hosts the long-running science podcast *The Curious Cases of Rutherford & Fry* with the BBC.

Adam Rutherford is an award-winning writer, broadcaster and geneticist at University College London. His books include *A Brief History of Everyone Who Ever Lived*, *The Book of Humans* and the *Sunday Times* bestseller *How to Argue with a Racist*. He has written and presented numerous documentaries for BBC radio and television, including *Inside Science* and *The Cell*. Adam has also worked as a science advisor on many films, including the Oscar-winning *Ex Machina* (2015) and *Annihilation* (2018).